Research Report

Fit for Duty?

Evaluating the Physical Fitness Requirements of Battlefield Airmen

Sean Robson, Maria C. Lytell, Carra S. Sims, Stephanie Pezard, Thomas Manacapilli, Amanda Anderson, Therese Bohusch, Abigail Haddad

RAND Project AIR FORCE

Prepared for the United States Air Force
Approved for public release; distribution unlimited

RAND CORPORATION

For more information on this publication, visit www.rand.org/t/RR618

Library of Congress Cataloging-in-Publication Data is available for this publication.
ISBN: 978-0-8330-8805-5

Published by the RAND Corporation, Santa Monica, Calif.

Foreword

The data described in this report were collected from 2011 to 2012 and provided the foundation for ongoing work conducted by the Air Force and RAND. Specifically, the results describing the physical requirements of Battlefield Airmen in this report have since been updated in a more recent job analysis conducted through a collaborative effort by the Air Force and RAND. The results from the updated job analysis will serve as the primary basis for conducting validation studies, which will provide the data necessary for recommending gender-neutral, occupationally relevant tests and standards for Battlefield Airmen.

Preface

Military occupations can be physically demanding, yet few attempts have been made to determine the physical readiness of today's Airmen to perform their jobs. Although the Air Force conducts fitness testing of all its Airmen, these tests and standards are not based on validated job requirements. Therefore, it is not clear how well an individual must perform on a given test (e.g., sit-ups) to be capable of performing a specific job, or if such a test even measures an ability required by the job. Recognizing the importance of measuring the physical readiness of Airmen, Air Education and Training Command (AETC) and Air Force Special Operations Command (AFSOC) asked RAND to provide a methodology to establish physical fitness requirements.

To demonstrate this methodology, we conducted an analysis of the physical demands and abilities of four occupational specialties (Combat Controller, Pararescue, Special Operations Weather Team, and Tactical Air Control Party). First, we identified the physical tasks required by each occupational specialty, using existing documents and eliciting feedback from subject-matter experts. Second, we narrowed the list of physical tasks to those most critical by conducting focus groups and interviews with operators in each career field. Third, we conducted an analysis of the physical abilities and movements required to perform the most physically demanding tasks in each specialty. These abilities provided the foundation for the fourth step, identifying valid and reliable fitness tests and measures.

This report describes research that extended over 12 months, from October 2011 through September 2012. The research was conducted within the Manpower, Personnel, and Training Program of RAND Project AIR FORCE.

This report should be of interest to Air Force leaders and staff concerned with standards to maintain the physical readiness of Airmen who perform physically demanding tasks as part of their occupational specialty.

RAND Project Air Force

RAND Project AIR FORCE (PAF), a division of the RAND Corporation, is the U.S. Air Force's federally funded research and development center for studies and analyses. PAF provides the Air Force with independent analyses of policy alternatives affecting the development, employment, combat readiness, and support of current and future aerospace forces. Research is conducted in four programs: Force Modernization and Employment; Manpower, Personnel, and Training; Resource Management; and Strategy and Doctrine. The research reported here was prepared under contract FA7014-06-C-0001.

Additional information about PAF is available on our website:
http://www.rand.org/paf/.

This report documents work originally shared with the U.S. Air Force on November 13, 2012. The draft report, issued on July 9, 2013, was scrutinized by U.S. Air Force subject-matter experts and formal peer reviewers.

Contents

Figures

Tables

Summary

The U.S. military requires service members who are physically capable of performing the many demanding tasks associated with their duties, which vary considerably by occupational specialty. One way the military accomplishes its goal of having physically capable personnel is by having them take periodic tests that assess physical fitness. Generally, these assessments provide a useful gauge of overall physical well-being, but they are not based on job requirements. Thus, the tests do not allow the military to determine whether personnel have the physical capabilities to carry out the specific tasks required by their occupational specialties.

Physical fitness standards may serve a wide range of goals, including improving general well-being, boosting unit morale, increasing productivity, reducing injuries and lost workdays, and eliminating stress. Depending on the goal, the standards could be the same for everyone or applied differentially, e.g., by age or sex. Standards developed for specific occupational tasks would be applied to all who perform those tasks.

Fitness standards also differ according to their type. Organizations often use two types of standards: norm-referenced and criterion-referenced. Norm-referenced standards reflect an individual's relative standing on a test relative to some referent group. For example, an individual who did 60 sit-ups in two minutes might be ranked in the 50th percentile of a referent group such as all soldiers between the ages of 18 and 21 years. Norm-referenced standards are essentially arbitrary and have no inherent meaning; they do not indicate whether an individual is healthy, can perform assigned duties, or deploy to combat. In contrast, criterion-referenced standards statistically link test scores with important outcomes or criteria. In the example above, criterion-referenced standards might link the ability to do 60 sit-ups in two minutes with the likelihood of developing heart disease.

Purpose and Approach of the Research

The U.S. Air Force has proposed a two-tiered approach to distinguish between fitness standards. Tier I standards are designed to reduce health risks and foster an overall fitness culture within the Air Force. Tier II standards are specific to Air Force Specialty Codes (AFSCs) and are intended to ensure that an individual is able to perform the physical tasks and duties required by his or her job. The Air Force asked researchers from RAND Project AIR FORCE (PAF) to demonstrate how the Air Force could establish Tier II fitness standards for physically demanding occupational specialties. We selected four AFSCs to study: Combat Controller (CCT), Pararescue (PJ), Special Operations Weather Team (SOWT), and Tactical Air Control Party

(TACP).[1] Airmen in these specialties are collectively referred to as Battlefield Airmen, and these specialties were selected in coordination with the Air Force because successful performance of the tasks associated with these specialties is believed to require high levels of physical fitness and ability. To help the Air Force develop Tier II fitness standards for Airmen, we set out to answer the following three research questions:

- What methods could be used to identify physically demanding tasks performed by Airmen?
- What methods are available for identifying the physical demands of occupational specialties?
- How can the Air Force use information about the physical tasks and demands to establish Tier II fitness standards?

To answer these questions, we employed a variety of methods. Specifically, we reviewed the methods used to establish physical requirements by different military organizations, the research literature on the different methods available for identifying physical fitness requirements, and documents and reports that contained specific tasks performed by Airmen in the four target specialties. Documents and reports included Occupational Analysis Reports (OARs), Career Field Education and Training Plans, Air Force Instructions, Air Force Technical Training Publications, and Army field manuals. After we developed an initial list of physically demanding tasks for each specialty, we vetted the tasks in focus groups and interviews, and we made efforts to observe Airmen performing the tasks during training. Interviews identified examples of critical physically demanding events that could significantly contribute to successful or unsuccessful performance. The outcome of these efforts was a list, by occupational skill, of the most important physically demanding tasks, or what we call "critical physical tasks" (CPTs).

Having identified CPTs, the next step was to link the tasks to the physical abilities needed to carry them out. This step included an analysis of movement patterns, such as balancing, carrying, lifting, climbing, and the physical abilities associated with the movement patterns, such as, muscular strength and endurance, cardiovascular endurance, and coordination, and agility.

The last step was to identify tests that could measure the required physical abilities. This identification involved both a literature review and an analysis of tests that have been validated with job performance.

Results

The physical abilities identified as critical for each of the occupational skills include the following:

- muscular strength

[1] Although we did not specifically include officers in our study, the results may be relevant to the following associated officer career fields: Combat Rescue Officer, Special Tactics Officer, and Air Liaisons Officer.

- muscular endurance
- cardiovascular endurance
- anaerobic power (ability to do high-intensity, short duration activity)
- equilibrium
- flexibility
- coordination and agility.

The radar chart in Figure S.1 shows the relative importance of the physical abilities for the four specialties as determined by focus groups and interviews. Lines closer to the center of the diagram indicate that fewer CPTs in that specialty require a particular physical ability, and those closer to the outer edge indicate that a greater number of physically demanding tasks are required for the specialty. The figure shows a high demand (more than 80 percent of the CPTs) across the four specialities for strength and muscular and cardiovascular endurance, followed by agility, anaeroboic power, and equilibrium. We found flexibility to be less critically important, only being required for slightly more than 25 percent of CPTs. Although there is considerable overlap in the physical abilities required for these specialties, further research is needed to identify which physical ability tests will predict operator performance in each occupational specialty.

Figure S.1. The Relative Importance of Battlefield Airman Physical Abilities

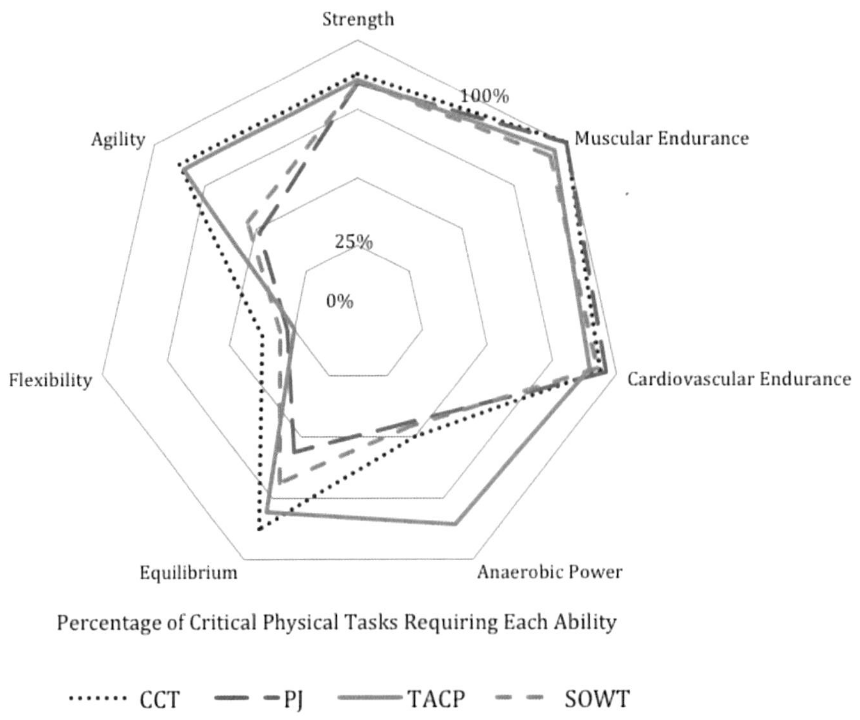

Recommendations for Developing Operationally Relevant Fitness Tests

The results of the analysis lead to one overarching recommendation and three specific recommendations.

Overarching Recommendation: Conduct a Validation Study to Establish Tier II Standards

The research reported here is a starting point. A critical next step for the Air Force is to engage in a systematic program of research and development to produce valid and reliable Tier II tests and standards to (a) ensure that tests measure important physical abilities required for successful mission or job performance, (b) ensure that performance on tests is a good indicator of mission or job performance, and (c) identify minimum test standards that are associated with acceptable mission or job performance. In selecting tests, particular attention should be paid to test reliability, cost and ease of administration and implementation, and, most important, coverage of the important physical abilities and tasks performed by operators.

Once tests are selected, the second step is to establish or develop appropriate job and training performance measures. This step will require additional work and collaboration with career field managers, squadron commanders, and subject-matter experts (SMEs). Determining what constitutes success is difficult, and we recommend developing behaviorally based performance evaluation scales for each Battlefield Airman specialty. For example, a behavioral observation scale (BOS) allows raters who are familiar with operators' performance to identify the frequency with which physically demanding tasks are performed effectively. For example, a rater may evaluate how often an operator keeps up with his team on overland movements or carries others team members' gear when they are fatigued. BOSs are reliable and effective methods for measuring performance. Once BOSs are developed, the next step is for SMEs to define the minimally acceptable level on the BOS. Final steps include analyzing the relationship between test scores and performance and establishing minimum scores.

Specific Recommendations

Use Multiple Tests to Measure Each Physical Ability

We recommend using at least two tests to measure each ability. Tests may include a combination of basic fitness tests and simulations, and can be integrated with the current Air Force Special Operations Command (AFSOC)/PJ operator tests and with tests conducted by strength and conditioning coaches assigned to the different squadrons. However, any test considered for validation should follow a strict protocol to ensure consistent administration, scoring, and reporting. Although flexibility was identified as an important ability for each Battlefield Airman specialty, we do not expect flexibility to relate strongly to performance. In fact, the quantitative review suggests that flexibility is among the weaker indicators of performance. However, poor flexibility has been cited as a potential indicator of injury risk; therefore, the Air Force may consider evaluating the potential benefits of a flexibility test by using injuries as the criteria.

Use Simulations to Offset Body Weight Bias of Basic Fitness Tests

Many basic fitness tests (e.g., pull-ups, three-mile run) are potentially biased in favor of smaller, leaner operators. We recommend integrating a job simulation that samples CPTs. Although specific simulations can be developed for each career field, we recommend developing a simulation that includes CPTs that are shared among all Battlefield Airman specialties. For example, CPTs shared across specialties include a march carrying a rucksack and carrying or dragging a casualty.

Test Alternative Methods for Setting Standards

After operators have been scored on the various tests and simulations, steps can be taken to identify the optimal combination of tests needed to determine physical readiness. We recommend using a compensatory model, which allows stronger performance on one or more tests to make up for slightly weaker performance on other tests. Although allowing weaker performance on some tests may seem counterproductive to reaching physical readiness goals, a compensatory model ensures that operators have the right combination of physical abilities to perform CPTs.

Final Thoughts

This report outlines the steps and provides an example, using the four enlisted Battlefield Airman specialties, of how to identify job-specific physical demands and the physical abilities needed to perform those tasks. Although this study focused on occupations closed to female Airmen at the time, the approach we took for developing occupationally relevant fitness standards and our recommendations for a validation study are relevant to the issue of women entering previously closed occupations. Recent changes in U.S. Department of Defense policy excluding women from certain assignments and specialties add urgency to the need for the services, including the Air Force, to establish appropriate gender-neutral standards for military occupations. Setting fitness standards that are tied to physical job performance is a key element to setting gender-neutral standards. Our study and recommendations can therefore inform efforts that the Air Force can take to address changes to the combat exclusion policy.

Acknowledgments

Many people contributed to the success of this project. We particularly recognize the efforts and support of Maj Gen Timothy Zadalis, Neal Baumgartner, Lt Col Scott Walter, Maj Becky Carter, and Lt Col Steve Goodman for helping coordinate and facilitate data collection, briefings, and meetings. We thank the CCT, PJ, SOWT, and TACP career field managers and training pipeline managers who met with us at the early stages of the project to provide guidance. We also thank the many others (e.g., squadron commanders, noncommissioned officers) who helped to support our data collection efforts, organize observations, and facilitate focus groups and interviews.

We especially thank the subject-matter experts Joseph Knapik, Marilyn Sharp, Peter Lisman, Scott Lephart, and Paul Davis, who provided feedback on the importance of physical ability dimensions; we recognize and appreciate the time it took to complete this effort. Scott Lephart was also instrumental in identifying advantages and disadvantages of various physical fitness tests.

We'd also like to thank program leaders from the U.S. Army's THOR3 program and Patrick Gagnon and Michael Spivock from Canada's National Defence for sharing information on their program's goals, challenges, and future directions. We are also grateful to all the Battlefield Airmen who participated in the interviews and focus groups and provided assistance in scoping the physical tasks for data collection.

Thanks to Drew Reinert and Chris Knerl for inviting our team to participate in the Human Performance Summit and for providing feedback and guidance on our efforts. We'd also like to acknowledge Christina Dozier and Jennifer Walters for providing data entry support, and Jerry Sollinger for assistance in preparing the summary and providing feedback on the report.

Abbreviations

AETC	Air Education and Training Command
AFSC	Air Force Specialty Code
AFSOC	Air Force Special Operations Command
ASOS	air support operations squadron
BOS	behavioral observation scale
CCT	Combat Controller
CWS	Combat Weather Squadron
CDF	Canadian Defence Forces
OAR	Occupational Analysis Report
PJ	Pararescue
RAMZ	Rigging, Alternate Method Zodiac
RQS	Rescue Squadron
SF	Special Forces
SME	subject-matter expert
SNCO	senior noncommissioned officer
SOWT	Special Operations Weather Team
STS	Special Tactics Squadron
TACP	Tactical Air Control Party

1. Introduction: Establishing a Need for Fitness Standards

Entry and Annual Fitness Standards

Many organizations, including public safety departments (e.g., fire, police) and military services, need to ensure that employees responsible for performing physically demanding tasks and duties have the necessary abilities to perform their jobs safely and effectively. Although many organizations have established entry standards to screen out those who would be unlikely to perform such tasks successfully, far fewer organizations have developed and implemented *annual* physical fitness standards based on *job requirements*. Although most public safety departments use entry-level screening tests, opposition from labor unions often prevents widespread use of annual fitness standards.

Unlike most public safety departments, the U.S. military services have annual fitness requirements. The tests and standards used to meet fitness requirements are generally not based on job requirements but rather focus on the general goal of maintaining a healthy force. Although maintaining good health is one important component of physical readiness, it does not guarantee that personnel can effectively perform the physically demanding tasks associated with their occupational specialties. One exception to this finding is the U.S. Marine Corps, which recently implemented the Marines Combat Fitness Test. Despite some limitations of this test, the Marine Corps' objective in using the test is to ensure that marines maintain sufficient levels of fitness to execute physically demanding job or mission tasks.

Differences Among Physical Fitness Standards

An organization's fitness standards can vary along several dimensions, including (a) purpose for developing standards, (b) type of standard, and (c) the specificity or target population.

On the surface, the purpose of developing physical fitness standards might seem clear. However, mandating specific fitness levels may serve a wide range of goals, such as improved health, unit morale, productivity, or readiness, or decreased injuries, lost workdays, or reduced stress. Depending on the goal, fitness levels might be the same for all military personnel, or standards might vary by age, gender, occupational specialty, and rank. For example, a fitness standard developed with the goal of improving overall health may determine that minimally acceptable fitness levels could be higher for younger, male personnel. In contrast, fitness standards developed with the goal of ensuring physical readiness to perform occupationally relevant, physically demanding tasks may use one standard for all personnel expected to perform those physically demanding tasks.

In addition to having different purposes, fitness standards have different types. Two types of standards are frequently used in organizations: (a) norm-referenced and (b) criterion-referenced standards. Norm-referenced standards reflect an individual's relative standing on a test compared with some referent group. For example, an Army soldier performing 60 sit-ups in two minutes may score at the 50th percentile among all male soldiers. The percentile provides information only about relative standing on the test; it has no inherent meaning. Consequently, norm-referenced standards are arbitrary; scoring at the 50th percentile does not indicate whether the soldier has good health, can perform his or her job, or is physically ready to deploy. To determine whether personnel have good health, are able to perform physical aspects of their jobs, or are physically ready to deploy, criterion-referenced standards would need to be established by statistically linking test scores with important outcomes or criteria (e.g., cardiovascular disease). For example, scores on a test involving sit-ups can be statistically analyzed to determine whether test performance predicts the risk of developing cardiovascular disease.

Finally, standards can vary by their level of specificity. For example, fitness tests and standards may be applied to all soldiers (e.g., a deployment readiness test). Additional standards could also be developed to enhance readiness by specifying the physical demands associated with particular units, squadrons, or occupational specialties. In general, an increase in the specificity of standards should be associated with higher levels of physical readiness.

Tier I and Tier II Fitness Standards

The U.S. Air Force has proposed a tiered model for distinguishing between physical fitness standards. Tier I standards are designed to reduce health risks and promote an overall fitness culture within the Air Force. Tier II standards are specific to Air Force Specialty Codes (AFSCs) and are intended to ensure that Airmen are ready to perform the physical tasks and duties of their AFSCs (see Table 1.1). The current Tier I standards were implemented in 2010 and provide age- and gender-graded standards on several fitness measures, including a 1.5-mile run. Minimum standards are based on observed statistical relationships between fitness scores and specific health risks (e.g., cardiovascular disease). To our knowledge, the Air Force is the only U.S. military service to use criterion-referenced cutoff scores for its fitness standards. The alternative method, previously used by the Air Force, is to use norm-referenced cutoff scores that provide the relative standing on the test compared with some referent group (e.g., percentiles). Although some have argued that the Tier I fitness should be modified to better predict the ability to perform combat-related tasks (Worden and White, 2012), these changes could be captured in additional standards (i.e., Tier II) that aim to ensure that military personnel are physically capable of performing essential mission tasks. Some of these physically demanding tasks may be required of all personnel, whereas other tasks are specific to particular occupational specialties.

Table 1.1. Primary Objectives of Tier I and Tier II Fitness Standards

Objective	Tier I Fitness	Tier II Fitness
Purpose	Reduce health risks Promote culture of fitness	Physical readiness
Target population	All Airmen	Physically demanding AFSCs
Tests	1.5-mile run, abdominal circumference, push-ups, sit-ups	Dependent on job demands
Standard	Based on projected health outcomes Varies by age and gender	Based on projected job performance outcomes Same standard for those performing same job duties

Purpose of Current Study

The purpose of this project study was to identify the physical ability requirements for a subset of occupational specialties in the Air Force. In particular, we established scientific methods to (a) identify important, physically demanding tasks, which we call "critical physical tasks" (CPTs), and to (b) identify the physical abilities required to perform these CPTs. These steps form the basis of a job analysis and can be replicated for any physically demanding career specialty. They also provide the foundation for future validation studies. Such validation studies need to be conducted to establish scientifically based standards for ensuring that individuals assigned to a particular occupational specialty have the necessary physical abilities to perform their jobs safely and effectively. Additional details on conducting validation studies are provided in Chapter Two.

In the current project, we conducted a physical job analysis for each of four physically demanding AFSCs: Combat Controller (CCT), Pararescue (PJ), Special Operations Weather Team (SOWT), and Tactical Air Control Party (TACP).[1] These occupational specialties are called "Battlefield Airman" specialties and were selected in coordination with the Air Force because Battlefield Airmen are believed to require high levels of physical fitness and ability to successfully perform their job tasks. Future studies would need to examine the physical requirements of other AFSCs to identify their physical ability requirements.

Since Battlefield Airman specialties are physically demanding, the Air Force has implemented rigorous entry, training, and annual fitness standards for these specialties. The entry standard, known as the Physical Ability and Stamina Test (PAST), includes an assessment of performance on push-ups, sit-ups, pull-ups, a 1.5-mile run, a 500-meter swim, and a 2 x 25 meter underwater swim.[2] Trainees are expected to improve their physical fitness during training and must meet higher standards to graduate from technical training.

---

[1] Although we did not specifically include officers in our study, the results may be relevant to the following associated officer career fields: Combat Rescue Officer, Special Tactics Officer, and Air Liaisons Officer.

[2] TACPs do not take the PAST but do have to pass similar tests, with the exception of the swim tests. In addition, TACPs do not have an annual fitness test that spans the entire specialty.

Although these specialties also have annual standards (listed in Table 1.2), which approximate Tier II standards, they differ from Tier II standards in at least two ways. First, the annual standards are not based on a thorough analysis of physical demands to determine which abilities and tests are important for measuring physical readiness. Second, the annual standards provide slightly easier minimally acceptable scores for older operators. Rather than automatically set different standards by age, Tier II standards would be based on the physical tasks operators perform or are expected to perform. Unless older operators are exempt from performing physically demanding tasks, the same standards would be set for all age groups. A more thorough discussion of this issue is presented in Chapter Two.

Table 1.2. Annual Fitness Standards for CCT, SOWT, and PJ Occupational Specialties

Physical Fitness Evaluation Criteria									
	Calisthenics (minutes)				Three-Mile Run		1,500-Meter Swim		
Battlefield Airman Specialty	Push-up (2)	Sit-up (2)	Pull-up (2)	Points per Test	Time (min:sec)	Points	Time (min:sec)	Points
PJ, SOWT, CCT	85	100	16	100	20:00	200	26:00	200
	80	95	15	90	20:30	190	27:00	190
	75	90	14	80	21:00	180	28:00	180
	70	85	13	75	21:30	170	29:00	170
	65	80	12	70	22:00	160	30:00	160
	60	75	11	65	22:30	150	31:00	150
	55	70	10	60	23:00	140	32:00	140
	50	65	9	55	23:30	130	33:00	130
	45	60	8	50	24:00	120	34:00	120
	40	55	7	45				
	35	50	6	40				
PJ only					25:00	110	35:30	110
					27:00	100	36:00	100
CCT and SOWT					25:00	110	34:30	110
					26:00	100	35:00	100

RATING SCALE		
	PJ	SOWT/CCT
OUTSTANDING	565–700	565–685
EXCELLENT	435–560	435–560
SATISFACTORY	320–430	325–430

Although TACPs do not have annual fitness standards that span across the specialty, some efforts have been made to establish such standards. For example, we were informed that the 19th Air Support Operations Squadron (19 ASOS) has implemented similar tests to those in the PAST (i.e., push-ups, sit-ups, pull-ups, three-mile run) as well as a five-mile ruck,[3] to evaluate the physical fitness of their TACPs.

Existing Research on Battlefield Airmen

A few research reports and studies have examined physical fitness requirements of Battlefield Airmen. Some of these studies were designed to improve fitness training (Walker et al., 2010), while others identified characteristics of successful trainees (Kalns et al., 2011). For example, Walker et al. (2010) evaluated 109 CCTs who successfully completed the first phase of training. The results indicated that successful trainees had approximately 12 percent body fat and scored well above norms on VO_2max, a measure of cardiovascular endurance, and on tests for strength and power capacity. A separate study examining success in the TACP training pipeline demonstrated that successful trainees perform significantly better on physical fitness tests (i.e., crunches, run, pull-ups) compared with those who failed training (Kalns et al., 2011). While these studies are useful in understanding the requirements to succeed in training, their ability to generalize to operators is limited to the extent that training is not representative of all of the physical demands typically experienced by operators. An analysis of operator physical demands is necessary to identify the physical fitness levels required to set annual standards.

Organization of This Report

Chapter Two provides an overview of the major steps required to validate physical ability tests and establish occupationally relevant physical fitness standards. Chapter Three provides an overview of our methodology for identifying the physical demands and abilities required of Battlefield Airmen. More specifically, it presents the job analysis procedures we used to identify CPTs, the physical movements associated with them, and the relative importance of physical abilities (e.g., muscular endurance) for successfully performing the CPTs. Chapter Four presents the results from the job analyses and is organized by specialty (e.g., CCT). In each section of Chapter Four, we provide background information on each specialty, information on the research participants and locations used to collect data, and specific findings from focus groups, critical incident interviews, movement pattern analysis, and the physical ability analysis. Chapter Five summarizes next steps and specific recommendations for conducting a thorough validation study for each specialty.

---

[3] The five-mile ruck consists of a march while carrying a rucksack (called "rucking"), which must be performed in combat uniform and boots with no less than 50 lbs. of gear without water.

2. Validation and Setting Standards

Testing for work-related physical standards can be defined as "the evaluation of a worker's capacity to work without risk to their own or others' health and safety" (Serra et al., 2007). Such testing is key to recruiting and personnel selection to ensure that only individuals with the appropriate capacities are hired. However, testing can also help to ensure that personnel are physically ready to effectively perform their jobs while not being at risk for work-related injuries.[1]

Standards for physical readiness can be implemented in a variety of ways, including regular performance evaluations, testing for medical fitness with physiological measures (e.g., pulmonary capacity or heart condition), and occupationally relevant fitness testing, which tests the capability to perform critical or essential job tasks. Although regular performance evaluations can be used for jobs that involve physically demanding tasks that are both observable and performed with some regularity, tests are often needed to ensure the physical readiness of individuals to perform critical tasks that may not occur regularly or are difficult to observe.

These tests should be based on the physical demands required to safely and effectively perform important or essential tasks. Tests that are too remote from actual job demands raise two main concerns. First, such tests may fail to screen individuals for the capacities they actually need in their jobs, potentially increasing the risks or threats to these individuals or their team. Second, these tests may be challenged before the courts if those who fail them can prove that such standards have no justification and are a basis for unlawful discrimination (Jackson, 1990).[2] It is therefore critical to conduct a thorough job analysis, which outlines the physical demands of job tasks on which such tests will be based. Tests should also focus on essential job functions rather than all job functions, for fear of losing some individuals who are otherwise capable of performing the essential job tasks—and increasing risks of litigation (Serra et al., 2007).[3]

Consequently, the purposes for setting fitness standards include a desire to maintain a workforce that is physically healthy, has few (if any) on-the-job injuries, and is capable of fighting the nation's wars. For Battlefield Airmen, capability means they can perform their missions successfully. Throughout this report, we provide information that will help the Air Force move toward physical performance standards for Battlefield Airmen.

In this chapter, we provide an overview of validation and standard setting (see Figure 2.1), as well as additional details about the theory, history, and alternative methods for validation. Since

---

[1] On the relationship between risk of injury and fitness level, see for instance Harwood, Rayson, and Neville (1999) and Knapik et al. (1993).

[2] See also Romines,1998, p. 10; Stanish and Campagna, 1999; and Shephard, 1990.

[3] See also Romines, 1998, p. 10.

validation is a critical step in developing relevant Tier II physical fitness standards, we integrate the results from our data collection with best practices to guide the Air Force in conducting a thorough validation for each Battlefield Airman specialty. Some of the data required for validation were collected as part of this study, but additional steps are needed to ensure that the fitness tests and standards selected for Battlefield Airmen are useful indicators of their physical readiness to perform their jobs.

Figure 2.1. Overview of Process For Fitness Test Validation

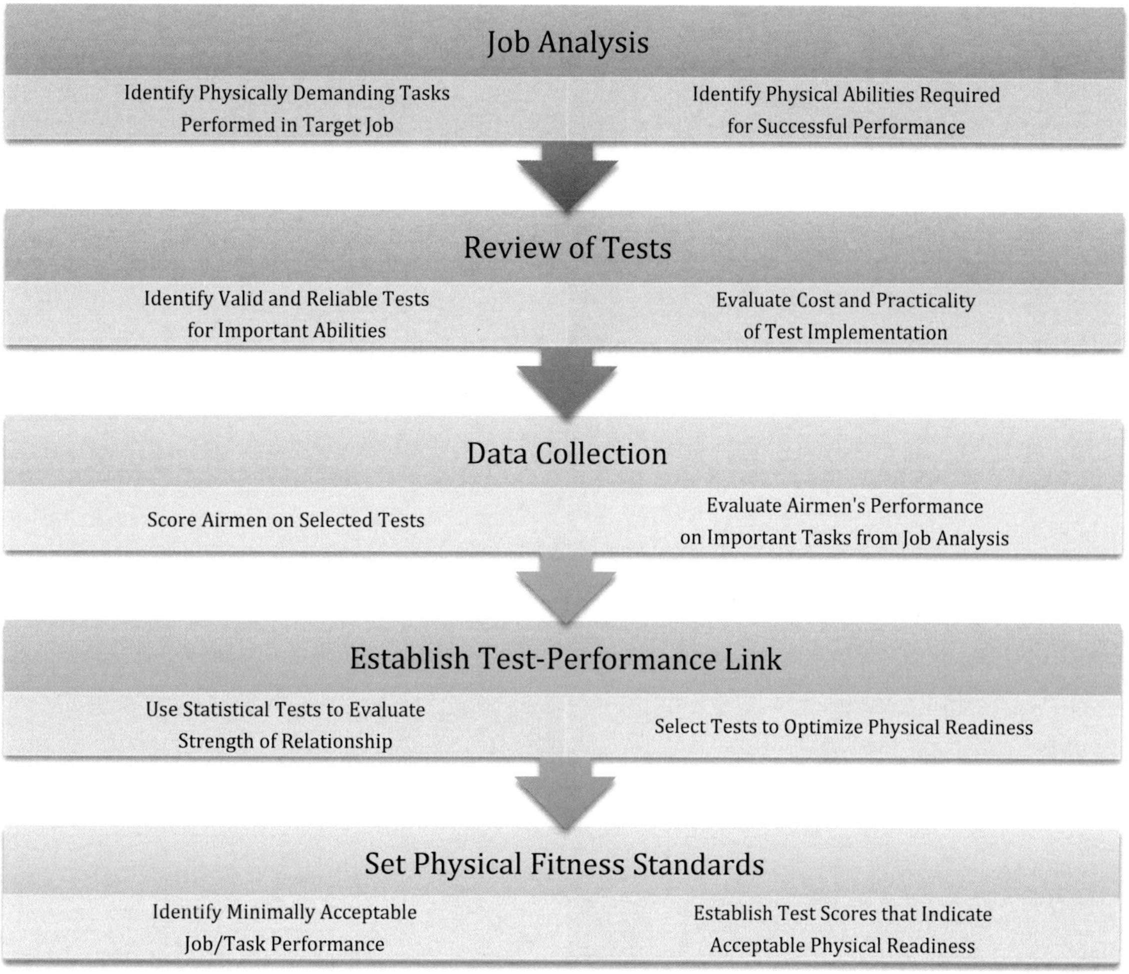

What Is Validation?

In education and employment, testing lies on a foundation of guidelines and legal precedent. These have been established primarily in the civilian domain but, as standard best practice, are also applicable to the military context. Regardless of whether an employer is concerned with potential litigation, the issues and remedies found in several sources, including legislation, case law, and professional guidelines, provide relevant recommendations on how to ensure that a test is usefully serving its purpose.

In general, these sources clearly indicate the importance of accumulating evidence to demonstrate that (a) the tests used are related to important job tasks, (b) the tests used measure important abilities required to perform critical job tasks, and (c) tests can discriminate between those Airmen who are physically ready and those who are not. Different strategies are needed to address each of these questions. In the sections below, we will outline the steps, data, and analyses needed to effectively address these questions.

What Job Tasks Are Important?

The foundation for developing effective Tier II standards begins with a job analysis to identify important job tasks (e.g., McPhail, 2007; Schmitt and Sinha, 2011). Using these important job tasks to guide the selection of tests helps provide evidence of *content validity*, which is evidence that a test represents or samples tasks relevant to the target job. Although sampling every task from a job can maximize content validity, this strategy is neither practical nor necessary to ensure physical readiness. The emphasis should be placed on using tests that measure or sample *critical* job tasks rather than trivial tasks. In designing the job analysis for Battlefield Airmen, we identified a critical task by using subject-matter experts (SMEs; i.e., Battlefield Airmen) to first discriminate more important from less important job tasks and second to discriminate high-intensity tasks (i.e., high physical effort) from low-intensity tasks (see Chapters Three and Four).

Who Should Identify the Abilities Needed to Perform the Job?

Different types of experts can be used to identify the required abilities for a job. Asking appropriate SMEs to make the relevant linkages between tasks and required knowledge, skills, and abilities is a traditional and well-supported part of job analysis (e.g., Williams and Crafts, 1997). In this study, we selected SMEs (i.e., exercise physiologists and industrial/organizational psychologists) who were familiar with both military tasks and physical ability measurement (see Chapters Three and Four). These experts provided ratings of the physical abilities required to perform the CPTs.[4]

Incumbents and/or their supervisors could also be used to make valid and reliable ratings of the abilities required to perform job tasks. However, the additional time demands that would be required of Battlefield Airmen did not make this a viable option.

[4] Subject-matter expert ratings are a common, defensible approach for establishing the physical requirements of a job. These ratings can be validated with direct measures of physical ability requirements for specific tasks (see Appendix A for a brief review of job analytic approaches).

How Should the Abilities Be Measured?

Physical ability measurement can be described by two broad approaches: (a) standard or basic fitness tests to assess each physical ability and (b) job simulations, which sample important physically demanding tasks (Gebhardt and Baker, 2010; Jackson, 2000). Fitness tests typically measure only one ability (Jackson, 2000; see also Knapik, 2004), and so a battery of tests are generally needed to assess multiple physical abilities. Certain tests represent the gold standard for a type of physical ability. For example, for cardiovascular endurance, the gold standard of physiological validity is a direct test of VO_2max. That is, there is general professional agreement that this test is a measure of cardiovascular endurance (Knapick, 2004). However, it is often far more feasible and practical to use different physical ability tests that have strong relationships with the gold standard measure, such as a timed run over a certain distance.

Another way to assess required physical abilities is through job simulations (Gebhardt and Baker, 2010; Hodgdon and Jackson, 2000; Jackson, 2000). These typically involve replication of various CPTs from the job (e.g., a ruck march). Because job simulations directly mimic required tasks, they have a direct connection to content validity and are considered more "face valid"—lay observers and test-takers can more clearly see how the simulation test (i.e., simulation) relates to the job. Although not an essential aspect of constructing a test or assessment battery, face validity can be helpful in enabling the test-takers to see the fairness of the test and enhance their buy-in for the testing process (Gilliland, 1993). Furthermore, the face validity of a test could also help facilitate regular training. Operators who do not perceive a strong connection between the test and the job may be more likely to cram in training sessions a few weeks before the test just so they can pass. That is, they see the test as another administrative hurdle rather than an important source of feedback for their physical readiness.

Disadvantages of using job simulations include a less-controlled setting than a typical physical ability test, which may increase the chances of injury (Gebhardt and Baker, 2010; Henderson, Berry, and Mattick, 2007; Jackson, 2000). Although these types of tests are quite common for law enforcement and public safety professionals and have demonstrated high validity for those types of jobs (Arvey, Nutting, and Landon, 1992; Gebhardt and Baker, 2010; Jackson, 2000), they can be expensive to develop if multiple simulations are needed. Therefore, we recommend identifying CPTs and demands that are shared across jobs to reduce the number of simulations that may need to be developed.

How Can You Tell Which Tests Are Good Indicators of Physical Readiness?

Although prior work suggests which tests may be relevant and potentially useful for determining physical readiness, it is important to demonstrate how well these tests work for specific jobs within the Air Force. The relationship between tests and job performance can be accomplished by establishing criterion-related validity.

Criterion-related validity strategies assess the strength of the relationship between a measure or test that is used for selection, retention, promotion, etc., and some measure that is relevant for job performance (i.e., the "criterion"). Criterion-related validity is often seen as the key type of validity evidence for linking tests and assessments, as well as standards on those tests and assessments, to job performance.[5] Although different strategies can be used to measure the criterion, each source of performance information has strengths and weaknesses. For example, supervisory ratings, while quite often used (see, e.g., McPhail and Stelly, 2010; Wildman et al., 2011), are also subject to some known challenges, including a tendency on the part of supervisors to rate performance more leniently when ratings are used for job outcomes (Arvey and Murphy, 1998). Therefore, to obtain an accurate estimate of performance, we recommend collecting performance ratings specifically for the purpose of the validation study. That is, performance ratings should only provide information for the validation study and should not result in any personnel decisions or actions about the individual being rated.

Criterion-related validity can be accomplished using either a concurrent or predictive strategy. In *concurrent* validation, a sample of job incumbents is given a test, and the strength of the relationship between that test and their current job performance is examined. In terms of the selection context, setting an appropriate cutoff or standard can sometimes be challenging when using concurrent validation, because all of the incumbents may be very experienced and excellent at their job and employ skills and training to accomplish their work that may not be found in a sample of applicants (e.g., McPhail, 2007).

In predictive validation, test scores are collected over some amount of time (e.g., one year) before the measure of performance, and hence the strength of the relationship is indicative of how well the measure predicts later performance.[6] Although both concurrent and predictive approaches are acceptable, a concurrent strategy is more practical and can result in a more timely analysis. Furthermore, a predictive strategy may provide an unreliable estimate of the usefulness of a test, since physical abilities can change as a function of injuries not related to fitness (e.g., accidents) and training over time.

Establishing Standards

In selection and standard-setting contexts, it is important to set the cutoff score, or minimum acceptable performance on an assessment tool, at the level of a minimally acceptable performer.

---

[5] Recall that job performance may be defined and measured in various ways for the purposes of a given study, and may include injuries at work, supervisory ratings of job performance, and, in our case, of physical job performance. More general outcomes, such as work attendance, number of widgets produced, and speed of promotions, may also be used as evidence to justify a test or standard.

[6] Ideally, the test is given to the applicant pool and performance assessed on those retained as employees. To obtain the best measure of the strength of relationship, all applicants are hired and their later job performance assessed and related to the selection assessment. However, this ideal is rarely achieved in practice, and statistical corrections are employed to approximate the actual relationship.

That is, one sets the cutoff score corresponding to a worker whose performance is acceptable—but only just. Truxillo et al. (1996) notes that cutoff scores are not appropriate for all purposes, but they discuss situations in which cutoff scores are useful.

What Type of Standard Should Be Set?

Norm-referenced cutoff scores use an established score distribution to set cutoff scores. For example, scores may be set such that 70 percent of test-takers pass. These types of scores are used for the Air Force's Tier I fitness test. When the desired use of a cutoff score involves a determination of test-takers who are able to perform at a minimally acceptable level on the job, criterion references must be used. As this is the purpose of Tier II standards, a *criterion-referenced* approach is most appropriate and would be determined as part of a criterion validation strategy as recommended above.

How High Should the Standard(s) Be Set?

How can this recommendation be achieved? Although it is tempting to set the standards at a level that would virtually ensure mission success (i.e., to set them at a very high level rather than at the level of minimally acceptable performance), this is not appropriate, for several reasons. Although Battlefield Airmen may face extreme and unpredictable job demands, which require maximum performance—that is, peak performance levels—setting test standards at this level may be impractical and may be quite different from more typical performance (Dubois et al., 1993). The test standards should reflect the *minimum* level required for effective and safe performance. However, operators can and should be encouraged to train above these minimum standards to further optimize their performance.

In some cases, the Air Force may consider setting a standard at a higher level than can be met by some job incumbents, even though this results in fewer successful operators available for deployment. In general, it is tenable to set standards higher than current incumbents can meet if the goal is to improve the quality of the overall workforce, but otherwise can be a problematic strategy as it would be subject to legal challenge (Gebhardt and Baker, 2010; Gutman, 2012; Truxillo et al., 1996). However, if the Air Force determines that, in general, Battlefield Airmen are successfully meeting their mission requirements, there is no evident need to improve the quality of the workforce. Furthermore, setting standards too high could result in not having sufficient operators available to meet the mission, which is a critical practical constraint.

Setting standards too high can also lead to increased concerns about potential injury risk to the operators. When a standard is overly high or a test itself potentially injurious, the risks of testing can outweigh the benefits: rather than ensuring a workforce that is able to meet the challenges of a physically demanding job, an organization might be faced with a workforce that is disabled through testing and unable to perform their jobs. In some jobs, such as those of Battlefield Airmen, injury may be anticipated as part of engaging in combat; however, there is also an imperative to keep Airmen fit to be able *to engage* in combat. Testing and assessment, or training for testing and assessment, should be managed to avoid such injuries when possible.

Conclusion

The appropriate standard is one that is related to important job performance criteria based on the results of a validation study, and set at the level of the minimally successful performer. This minimum level provides the foundation for promoting further performance gains through other strategies (e.g., training). Performance above the minimum standard can and should be encouraged to the extent that greater mission success can be achieved by increasing fitness in one or more abilities. The bottom line is that minimum standards ensure that regular job performance needs will be met without undue risk of being overly selective or injurious.

3. Our Approach for Identifying Physical Requirements of Battlefield Airmen

To identify the physical requirements of the four Battlefield Airman occupational specialties in our study, we reviewed a variety of established scientific approaches (see Appendix A). Based on our review, we developed a two-step process that we outline in this chapter to identify the physically demanding tasks performed by operators and the abilities required to perform these tasks. That is, these steps involve conducting a job analysis to identify the physical requirements of each occupational specialty. These steps are part of the overall validation process described in the Chapter Two.

Step 1. Identify Critical Physical Tasks

This section provides the steps we took to identify CPTs. We defined a CPT as a task that requires at least a moderate amount of physical effort and is important to the specialty. Identifying the CPTs for each specialty provides the foundation for additional data collection about the physical demands associated with each of these tasks. Without a thorough understanding of which tasks are physically demanding, it would be very difficult to identify which physical abilities are needed and their relative importance to overall success as an operator. The specific procedures and criteria for selecting CPTs are presented in the sections below.

Preliminary Task List

We used several sources of information to identify physically demanding job tasks and duties, including meetings with SMEs, observations, and a review of Air Force documents and reports. After an initial meeting with the career field managers of the four specialties in the study, we reviewed several documents and reports that contained specific tasks Airmen in these specialties perform. These documents and reports included occupational analysis reports (OARs), career field education and training plans (CFETPs), Air Force instructions (AFIs), Air Force technical training publications (AFTTPs), and Army field manuals (FMs).[1] Using the OARs as our primary source of job and duty tasks for each career field, we established a preliminary list of physically demanding tasks by eliminating tasks that were clearly nonphysical (e.g., supervisor task of writing performance reports). This preliminary task list was then reviewed by training

---

[1] We reviewed Army field manuals because several elements of training for Battlefield Airmen rely on guidance based in these manuals. Also, Battlefield Airmen training includes courses conducted by the Army (e.g., parachute jump training at Fort Benning, Georgia).

pipeline managers and senior noncommissioned officers (SNCOs) to eliminate tasks that did not require some physical effort to complete and to identify any additional physically demanding tasks not covered in our preliminary list (see Table 3.1). This review process resulted in 47 to 69 physically demanding tasks for each career field. These tasks were further evaluated by operators in a series of focus groups designed to identify the importance, intensity, frequency, and duration of each task.

Table 3.1. Steps Used to Develop Preliminary Physical Task List

Evaluation Stage	Number of Tasks Reviewed			
	CCT	TACP	PJ	SOWT
Original List	379	460	767	597
Elimination of nonphysical tasks	125	164	481	148
SME review	67	69	47	67

Focus Groups

We conducted focus groups for each occupational specialty at several locations with representative Special Tactics Squadrons (STSs), a Rescue Squadron (RQS), a Combat Weather Squadron (CWS), and several ASOSs (see Table 3.2). The focus groups were used to administer a survey, which asked operators to rate four dimensions for each physical task: (a) importance, (b) frequency, (c) duration, and (d) intensity (see Appendix C for dimension definitions and scales). For importance, operators were asked to rate how important each task was to effective mission performance on a 5-point scale ranging from 1 "Not Important" to 5 "Crucial." For frequency, operators were asked to rate how often they performed each task in their missions over the past two years on a scale from 1 "Never Performed" (e.g., received training but never had to perform in a mission) to 5 "Always Performed" (e.g., on every mission, multiple times during a mission). To assess duration, operators were asked to rate how long it takes to perform each task before they could either rest or move on to a new task. Duration was rated on a 5-point scale ranging from 1 "0 to 2 minutes" to 5 "More than 2 hours." Finally, operators were asked to rate how much physical effort was required to successfully complete each task to measure intensity, which was rated on a 7-point scale ranging from 1 "Very, Very Light" to 7 "Very, Very Hard." This scale is an adaptation of Borg's perceived exertion scale (Borg, 1970) and has been shown to correlate well with metabolic costs required to perform tasks (Hogan and Fleishman, 1979). Additional details on the sample and the results from the focus groups are presented in Chapter Four.

Table 3.2. Focus Group Participants

CCT	PJ	SOWT	TACP
23rd STS	23rd STS	10th CWS[a]	23rd STS
21st STS	Air Education and Training Command (Instructors)		5th ASOS
	103rd RQS		9th ASOS
			14th ASOS
			19th ASOS
			20th ASOS

[a] Although active at the time of the study, the 10th CWS no longer exists.

Critical Incident Interviews

We conducted interviews with SNCOs[2] for each specialty to identify examples of critical physically demanding events that could significantly contribute to successful or unsuccessful performance. These semistructured interviews lasted approximately 45 minutes each and allowed interviewers to use one or more questions to elicit critical incidents:

- Think of a time when an Airman performed a physically demanding task beyond reasonable expectations.
- Think of a time when an Airman had difficulty completing a physically demanding task.
- Think of a time when an Airman successfully completed a task requiring strength, endurance, or quality of movement (i.e., balance, flexibility, coordination) that other Airmen would have had difficulty performing.
- Think of a time when an Airman did not have the strength, endurance, or quality of movement needed to successfully complete a task or mission.
- Think of a time when an Airman's fitness may have led to his or another team member's injury.
- Think of a time when an Airman's fitness may have increased the safety and security of others.

The critical incident interviews were coded by two independent raters to identify specific factors (physical tasks and abilities) that may affect both positive and negative operator performance. For each AFSC, the number of abilities and tasks were summed and averaged across the two raters. In addition to the physical abilities identified by our SMEs (see later section on our physical ability analysis), other abilities not identified by our SMEs were considered when specifically mentioned during the interviews. For example, core strength was identified as a critical ability in several interviews but not by SMEs.

[2] We also interviewed officers. However, the vast majority of interviews were with SNCOs, who we expected would be in a better position to identify critical events, as they have directly supervised the Airmen.

Observations

We also attempted to observe operators performing physically demanding tasks during training exercises. Although we made some limited observations for the PJ and TACP occupational specialties, observations were generally very difficult, for several reasons. For example, training exercises that would have been performed by CCTs and SOWTs during one of our visits were canceled because of bad weather. Due to these difficulties, we did not conduct any systematic analysis of tasks using information from observations of tasks being performed.

Although observation was not a successful source of information in this context, other occupational specialties may be more suitable for observation. For example, observations may provide useful information for specialties with physically demanding tasks that are performed routinely and in a controlled environment (e.g., maintainers working on aircraft in a hangar).

Selecting Critical Physical Tasks

Based on results from our focus group surveys, we selected a subset of tasks for each AFSC that represented the most physically demanding tasks (i.e., CPTs). We learned from our focus groups and critical incident interviews that Battlefield Airmen perform a variety of tasks, some of which are critical to the mission but might not be performed very often. As such, we did not rely on ratings of frequency of task performance to select CPTs. Instead, we selected tasks that were deemed by focus group respondents to be very important and require high levels of physical effort. We specifically chose tasks rated as 4.0 or higher on importance (i.e., "very important" or "crucial") and 4.0 or higher on intensity (i.e., "somewhat hard," "hard," "very hard," or "very, very hard"). We selected the 4.0 cut points on these two scales because they represented the minimum level at which a task is considered important (i.e., above the 3.0 level of "moderately important") and intense (i.e., above the 3.0 level of "light"). Although both scales were used to determine the cutoffs, emphasis was placed on the importance of each physical task to determine criticality to the specialty. We used the intensity scale primarily to screen out tasks that required minimal physical effort. Important tasks that require little physical effort may provide insight for other standards, but should not used to establish physical fitness requirements. We discuss the final lists of CPTs in Chapter Four.

Step 2. Link Critical Physical Tasks to Abilities

The second step in our process was to identify the different types of physical abilities that are needed to perform each of the CPTs. Identification of the underlying abilities provides the foundation for the selection of appropriate physical ability tests to ensure that operators are physically capable of performing the full range of CPTs. Although one or more CPTs may be performed in a team setting (e.g., four-person litter carry), the analyses in this section focused on the individual level of analysis. That is, the objective was to identify the specific movement patterns and physical abilities that would be required by each operator to be capable of performing each CPT. This approach allows the Air Force to identify physical ability and performance standards for operators within each occupational specialty.

Since the CPTs were not immediately transparent to the external SMEs (exercise physiologists and industrial-organizational psychologists) selected to identify the required abilities, we first conducted a movement pattern analysis to identify the specific movements (e.g., pulling) that were performed during the execution of each CPT.

Movement Pattern Analysis

A small group of SNCOs rated the extent to which various movement patterns were required to perform the CPTs. Each movement pattern was rated as 0 "not used," 1 "used to some extent," or 2 "used to a great extent." We averaged the ratings across SMEs within each AFSC. For a movement pattern to be coded as "extensively used," the average rating had to be between 1.25 and 2.00. To be coded as moderate, the average rating had to be between 0.50 and 1.24. Average ratings below 0.50 were coded as not required for successful task performance.

The movement patterns were adapted from the Dictionary of Occupational Titles. We eliminated some of the movement patterns from our list because they focused on fine psychomotor movements (e.g., using fingers) rather than movements of the major muscle groups. After reviewing the list, we also added "sprinting" to help differentiate between tasks requiring aerobic and anaerobic energy systems. The final list of 12 movement patterns and definitions is presented in Table 3.3. Additional details on the raters and the results are presented for each AFSC in Chapter Four.

Table 3.3. Movement Patterns and Definitions

Movement	Definition
Balancing	Maintaining body equilibrium to prevent falling. Includes balancing on level surfaces, balancing on uneven surfaces, balancing on ladders, and balancing on beam and scaffolding.
Bending/rotating	• *Forward Bending*: Bending the body downward and forward from a standing position by bending the spine at the hips and/or waist. The hips must be flexed more than 20 degrees and the knees are kept relatively straight (flexed no more than 35 degrees). • *Backward Bending*: To extend the back, backward (arching the back). • *Rotating:* Twisting the upper body (trunk) or bending the upper body to the side (lateral bending).
Carrying/lifting	• *Carrying:* Transporting an object over a distance through walking, usually holding the load in the hands or arms. Includes one-handed (using one hand or arm to carry the object) and two-handed (using both hands or arms to carry the object). • *Lifting:* Raising or lowering an object from one level to another. Involves primarily vertical displacement of the load but can also include a component of horizontal displacement as well. Can involve one or two-handed lifting and can occur either above waist or below waist. Includes one handed (using one hand or arm to raise or lower the object), two-handed (using both hands or arms to raise or lower the object), above-waist (lifting that occurs from the waist and above. Typically performed primarily with the strength of the arms, shoulders, and upper back) and below waist (lifting that occurs from the floor to approximately waist height. Typically performed primarily with the strength of the legs and low back.
Climbing	Ascending or descending stairs, ladders, scaffolding, poles, equipment, or other apparatus using the feet and legs, and/or hands and arms.
Crouching/squatting	• *Crouching*: Bending the body downward and forward by bending legs at the hips and knees with simultaneous forward bending of the spine. This is typically performed when working with material that is at or near the floor level. • *Squatting* includes positions where one knee is on the floor or both knees are off the floor.
Gripping/handling	Seizing, holding, grasping, turning, or working with hands; using the hands in such a fashion that the object being handled contacts the palm and fingers of the hand.
Pulling	Exerting force upon an object so that the object moves toward the force. Includes one handed (using one hand or arm to pull the object) and two handed (using both hands or arms to pull the object).
Pushing	Exerting force upon an object so that the object moves away from the force. Includes one handed (using one hand or arm to push the object) and two handed (using both hands or arms to push the object).
Running	To move swiftly on foot so that both feet leave the ground during each stride.
Sprinting	To move rapidly or at top speed for a brief period.
Standing/walking	• *Standing:* Remaining on one's feet in an upright position without walking. • *Walking:* Moving about on foot. Requires three consecutive steps to be considered walking. Includes walking on level surfaces (surfaces that are level and do not include ramps or uneven terrain), uneven surfaces (surfaces that include uneven terrain. Includes walking outside over grass, dirt, gravel, up and down curbs), and ramps/inclines (surfaces that include an incline of over 15 degrees).
Jumping	To spring up from the ground or down to a lower level using the foot and leg muscles.

Physical Ability Analysis

Exercise physiologists (n = 5) and industrial-organizational psychologists (n = 3) rated the importance of seven physical abilities (see Table 3.4, Gebhardt and Baker, 2010) on a 5-point scale ranging from 1 “not important” to a 5 “critically important.” To facilitate the rating process, each SME was provided with an Excel spreadsheet containing the physical abilities and definitions; the CPTs; average ratings on importance, frequency, duration, and intensity provided by the operators’ responses to focus group surveys; and a description of the movement patterns required to perform each CPT. The interrater reliability for all abilities was acceptable (intraclass correlation coefficients [ICCs] > 0.70) (Shrout and Fleiss, 1979).

Table 3.4. Physical Abilities and Definitions

Abilities	Definition
Muscular strength	Ability of the muscles to exert force to lift, push, pull, or hold objects. The magnitude of the force depends on the size of the muscles (cross-section).
Muscular endurance	Ability to exert force continuously over moderate to long time periods. The length of time a muscle can contract depends on the size of the muscles involved, the chemical composition of the muscle tissue, and the muscle fiber type (e.g., slow twitch).
Cardiovascular endurance	Ability of the respiratory and cardiovascular systems to provide oxygen continuously for medium- to high-intensity activities performed over a moderate time period (e.g., > 5 minutes).
Anaerobic power	Ability to perform high-intensity short-duration activities (e.g., 5–90 sec.) using stored energy (e.g., adenosine triphosphate).
Equilibrium	Ability to maintain the body’s center of mass over the base of support (e.g., feet) in the presence of outside forces (e.g., gravity, slipping on ice). Equilibrium involves maintaining and recovering a balanced position.
Flexibility	Ability to bend, stoop, rotate, and reach in all directions with the arms and legs through the range of motion at the joints (e.g., knee, shoulders). Flexibility is dependent on the extensibility of the ligaments, tendons, muscles, and skin.
Coordination and agility	Ability to perform motor activities in a proficient sequential pattern by using neurosensory cues such as change of direction.

4. Results: Critical Physical Tasks and Abilities

This chapter describes the results from the focus groups and SME ratings for each occupational specialty. First, it presents the CPTs, which met the criteria described in Chapter Three (i.e., average ratings above 4.0 on the importance and intensity scales). Second, it describes findings from the critical incident interviews, which were used to provide additional context regarding the physically demanding tasks performed, as well as results from the movement patterns analysis. Finally, the chapter presents the average SME ratings for the physical abilities (e.g., muscular strength) required to successfully perform each CPT.

Combat Controller

Background

Combat Controllers (CCTs) originated before the Air Force was a service, when Army pathfinders were sent to provide guidance for airdrops during World War II. They deploy into hostile or combat areas to establish airfields and assault zones. As qualified air traffic controllers, they also direct air traffic and provide command and control, as well as fire support. The majority also qualify as joint terminal attack controllers (JTACs). They deploy with special tactics teams and as such not only serve as air traffic controllers but are also expert at infiltration and exfiltration methods, including fast rope methods and free fall parachuting, rubber raft techniques, and survival evasion and resistance skills. As part of their air traffic control and communication duties, they also carry heavy communication equipment while on mission (U.S. Air Force, 2010c).

Focus Groups

Focus groups of CCTs were conducted with the 23rd and 21st Special Tactics Squadrons. A total of 57 operators participated in 11 focus groups. Participants rated 67 tasks according to their level of importance, frequency, duration, and intensity. The 67 tasks covered various physical activities, including movement under different conditions (e.g., at altitude), infiltration and exfiltration, and combat operations (see Appendix C for the full list of tasks). Based on participants' ratings of the task list, we identified 16 CPTs for CCTs. These tasks and the average importance rating on a 5-point scale ranging from 1 (not important) to 5 (crucial) are presented in Table 4.1 (see Appendix C for the full scale).

Table 4.1. CCT Critical Physical Tasks

Task	Importance
Carry casualties from immediate battle area to cover (e.g., use fireman's carry)	4.51
Drag casualties from immediate battle area to cover	4.78
Participate in surface operations in urban environments with ruck load	4.45
Participate in surface operations in deserts with ruck load	4.44
Participate in surface operations in extreme cold environments with ruck load	4.11
Participate in surface operations in jungles with ruck load	4.27
Participate in surface operations at high altitudes above 8,000 with ruck load	4.57
Perform dismounted offensive operations	4.64
Perform overland infiltration or exfiltration operations	4.57
Perform night Infiltration with follow-on mission	4.70
Perform operational fast rope drop into a landing zone, e.g., from a helicopter (with no equipment, with LBE, weapon, and helmet, and with full combat gear)	4.07
Conduct day and night tactical air control (TIC) missions	4.75
Perform immediate action drills (IADs)	4.63
Perform hand-to-hand combat	4.11
Perform night compass swim-scuba operations	4.09
Maneuver around, over, and under obstacles while in combat load	4.68

NOTE: Importance scale was rated on a scale ranging from 1 (not important) to 5 (crucial). The full scale is presented in Appendix C.

Critical Incident Interviews

We conducted interviews with eight CCT experts. Interviewees provided several examples of critical physically demanding incidents from their deployment missions and related experiences. Figures 4.1 and 4.2 present the most common types of abilities and tasks representing the critical incidents discussed during the interviews. As a reminder, the number of citations is an average of two independent raters who identified how many times a particular ability or task was mentioned. An ability or task was only counted once for each example provided, but an interviewee may have provided several examples of the same task or ability. Table 4.2 provides examples of the most common types of physical tasks that were described.

Figure 4.1. Abilities Most Often Cited in CCT Critical Incident Interviews

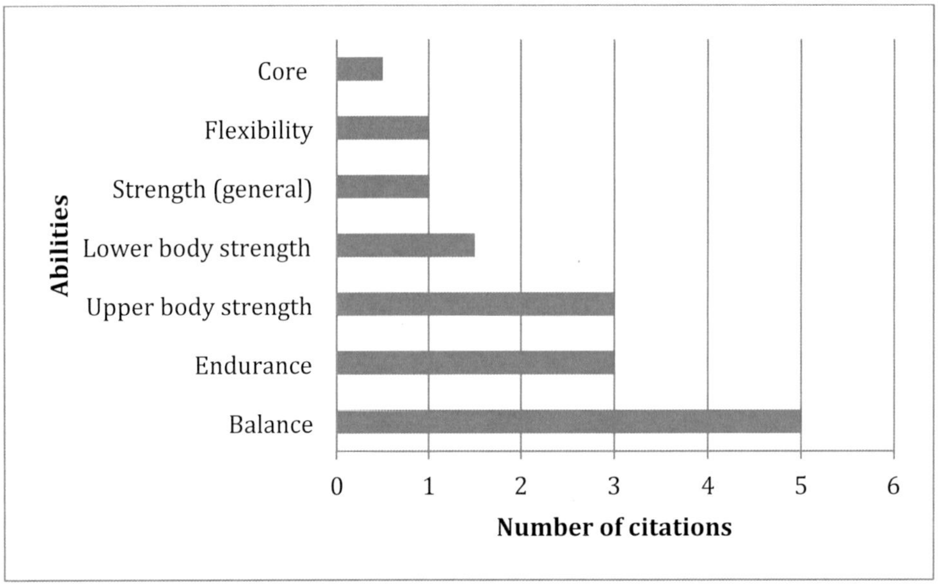

Figure 4.2. Tasks Most Often Cited in CCT Critical Incident Interviews

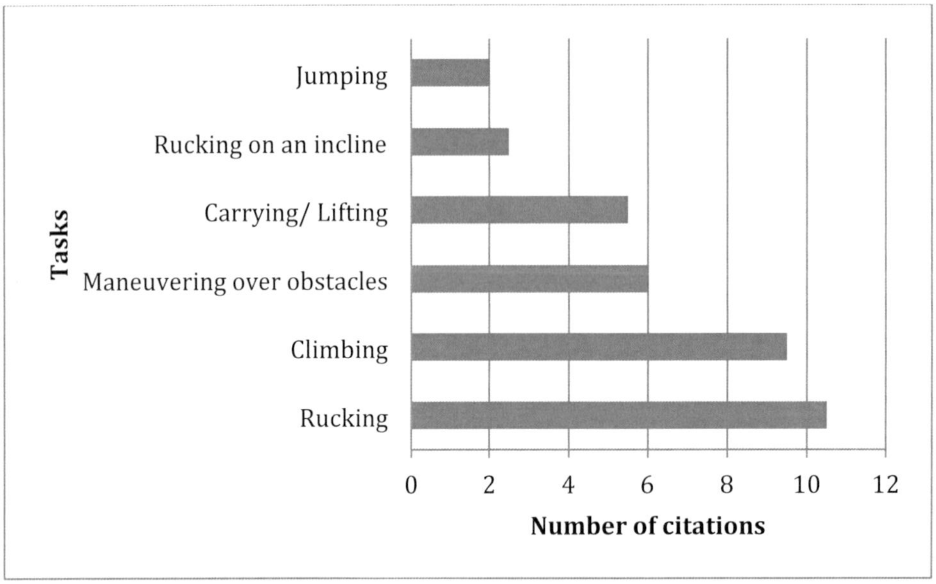

Table 4.2. Sample of Critical Incidents Described During CCT Critical Incident Interviews

Task	Number of citations	Examples
Rucking	10.5	• Having to stop for the slowest guy during a ruck march • Danger of taking the "easier" route during a ruck march
Climbing	9.5	• Climbing vertical walls of more than 10ft • Difficulty of working with forces who are not able to fast rope
Maneuvering over obstacles	6	• Getting over a building to get to cover • Hopping over objects
Carrying/lifting	5.5	• Jogging with weapons • Carrying injured people
Rucking on an incline	2.5	• Climbing up a hill • Rucking up a mountain in Afghanistan
Jumping	2	• Hopping over a 4ft object • Blowing one's knee when jumping off an helicopter

Movement Pattern Analysis

Ten CCT experts (SNCOs) rated the 16 CPTs according to primary movement patterns. As a reminder, we averaged the ratings and grouped them into three categories of "not used," "moderately used," and "extensively used" for the purpose of illustrating the relative frequency of each movement pattern for successful task performance. The average ratings across all CCT CPTs are depicted in Figure 4.3.

Gripping/handling and carrying/lifting are frequently used to perform CCT CPTs (i.e., rated as extensively used for at least 70 percent of CPTs). Standing/walking also received high ratings. Examination of the CCT CPTs in Table 4.1 shows that many involved surface operations under load, which illuminates the importance of both carrying/lifting (i.e., loads) and standing walking (i.e., rucking itself). Surface operations, hand-to-hand combat, casualty drags, and maneuvering around obstacles under load were also rated as requiring large amounts of gripping/handling. The full list of CPTs and average movement pattern ratings are provided in Appendix B.

Figure 4.3. Average Movement Pattern Ratings for CCT CPTs

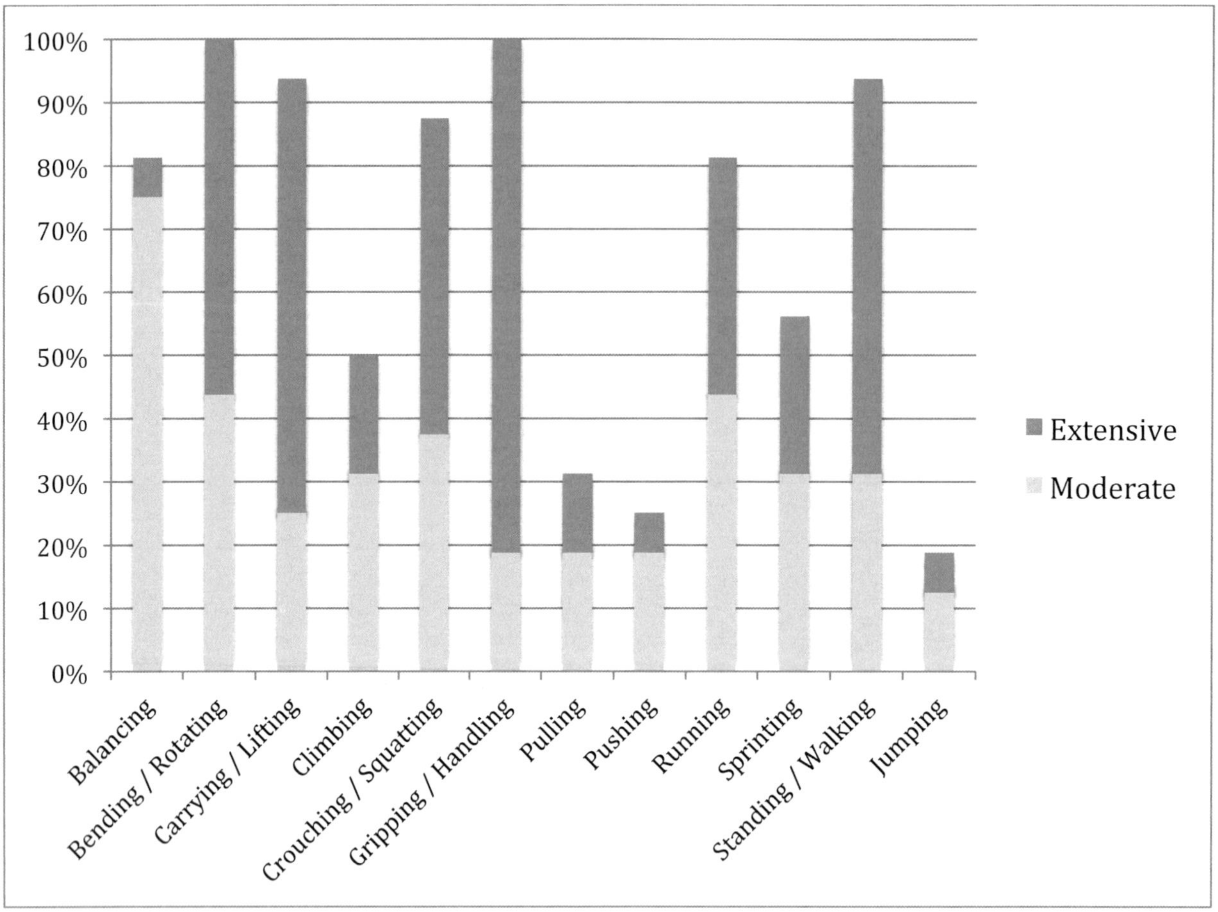

Physical Ability Analysis

The physical abilities required to perform CCT CPTs are depicted in Figure 4.4. Muscular and cardiovascular endurance were both required for over 90 percent of CPTs, as might be expected for operations that require participants to carry loads in varying conditions. Strength, equilibrium, and agility were also often required. Anaerobic power and flexibility are rated overall as less necessary. However, approximately 35–50 percent of CPTs require these abilities so they are needed to some extent.

Figure 4.4. Percentage of CCT CPTs Requiring Each Physical Ability

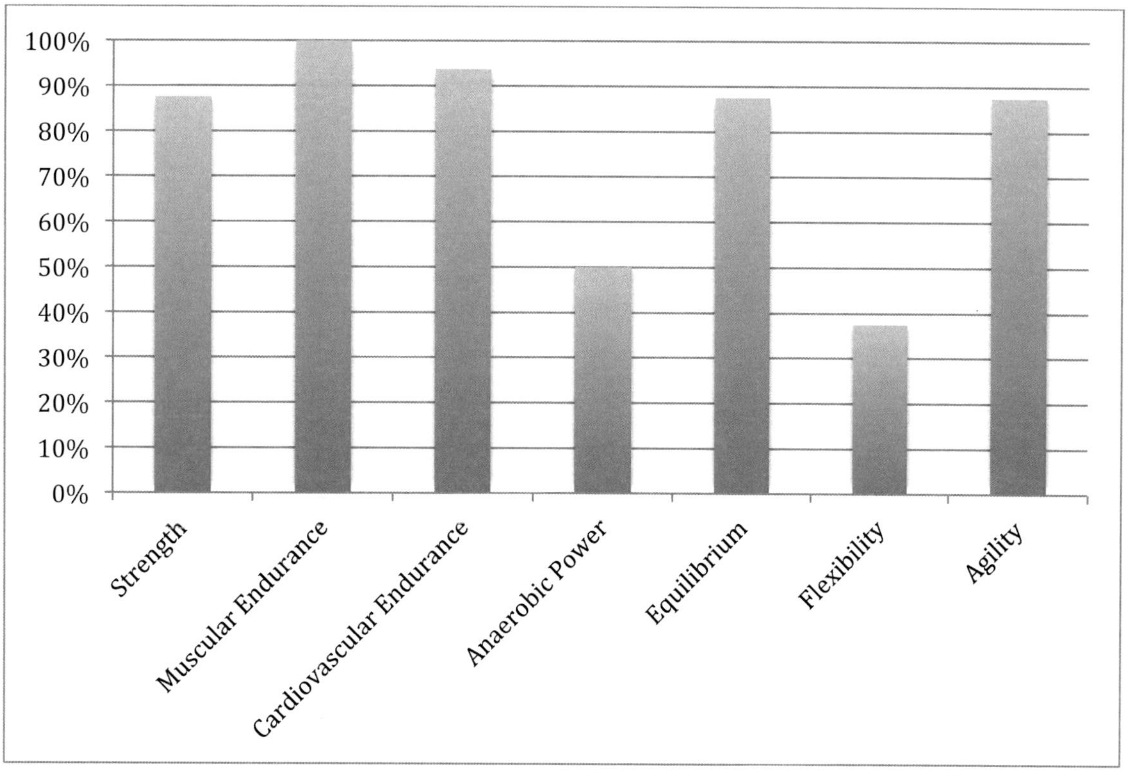

Pararescue

Background

Air Force pararescuemen, also known as PJs, are the only Department of Defense (DoD) elite combat forces specifically organized, trained, equipped, and postured to conduct full-spectrum personnel recovery (PR), including both conventional and unconventional combat rescue operations. Their primary mission is to rescue, recover, and return American or allied forces in times of danger or extreme duress. To accomplish this mission, PJs are qualified and trained in many areas, including advanced weapons and small unit tactics, airborne and military free fall, both High Altitude Low Opening (halo) and High Altitude High Opening (haho) parachute operations, combat dive, high angle/confined space rescue operations, small boat/vehicle craft utilization, rescue swimming, and battlefield trauma/paramedics (U.S. Air Force, 2010b).

Focus Groups

Focus groups of PJs were conducted with the 23rd Special Tactics Squadron, Air Education and Training Command (AETC) instructors, and the 103rd RQS. A total of 28 operators participated in four focus groups. Efforts to gain additional participation were hindered by deployments, training schedules, and designated leave time.

Focus group participants rated 47 tasks according to their level of importance, frequency, duration, and intensity. The 47 tasks covered various physical activities, from different rescue operations (e.g., high-angles rescue) to combat operations (see Appendix C for the full list of tasks). Based on participants' ratings of the 47 tasks, we identified 25 CPTs for PJs. These tasks and the average importance rating on a 5-point scale ranging from 1 (not important) to 5 (crucial) are presented in Table 4.3 (see Appendix C for the full scale).

Table 4.3. Pararescue PJ Critical Physical Tasks (CPTs)

Task	Average Importance
Perform immediate action drills and execute tactics to break enemy contact (i.e., I'm up, he sees me, I'm down, REPEAT)	4.93
Perform medical treatment of wounds, injuries, and illnesses including combat casualty care, triage of mass casualties, life saving techniques, crisis action team functions, and JMAU duties	4.93
Carry or lift survivors from aircraft or vehicle wreckage	4.89
Drag casualties from immediate battle area to cover	4.82
Employ and operate extrication devices and equipment (e.g., axe, sledgehammer, crowbar)	4.79
Carry patients from immediate battle area (e.g., fireman carry)	4.71
Lift, load and unload equipment or personnel in litters onto trailers, vehicles, or aircraft	4.64
Perform technical rescue and recovery for confined space or imminent collapse rescue operations	4.64
Perform downhill evacuation methods using low-angle evacuation, high-angle evacuation, buddy rappel, high-angle pick-off, improvised litters, suspension traverse method, rescue belay system, system know bypass	4.64
Maneuver body or litter over rocks or other obstructions during high angle rescue	4.57
Carry litter loaded with casualty for transfer or transload over flat or low-angle terrain	4.54
Perform water recoveries of personnel or materials	4.50
Climb rope ladder, caving ladder, or hook ladder to board aircraft or watercraft	4.50
Perform search and rescue (SAR) security team (SST) operations	4.50
Climb over walls and obstacles and crawl to stay in cover	4.46
Move over various terrain (e.g., jungle, mountain, desert, beach) with ruck load for extended surface operations, infiltration, or exfiltration	4.46
Prepare and execute open and closed-circuit dive operations including buddy rescue, underwater search and recovery, and dry-suit dive operations	4.46
Perform movement or reconnaissance over difficult and varied terrain (e.g., glacier, mountain travel)	4.43
Swim with fins to unpack Ducks and RAMZ	4.39
Perform rope ascents and descents using ascending devices, friction knots, and roped party climbs	4.39
Perform underwater searches, such as circle line, clump line, parallel, or SONAR	4.25
Perform standard and emergency aircraft operations such as emergency crash and egress	4.21
Haul lines for recovery of personnel or equipment	4.21
Perform actions to survive, evade, resist, and escape the enemy including constructing shelters	4.21

NOTE: Importance scale was rated on a scale ranging from 1 (not important) to 5 (crucial). The full scale is presented in Appendix C.

Critical Incident Interviews

We interviewed a total of six PJ NCOs and held one group interview with AETC instructors. Interviewees provided several examples of critical physically demanding events from their deployment missions. Figures 4.5 and 4.6 present the most common types of abilities and tasks representing the critical incidents discussed during the interviews. Table 4.4 provides examples of the most common types of physical tasks that were described.

Figure 4.5. Abilities Most Often Cited in PJ Critical Incident Interviews

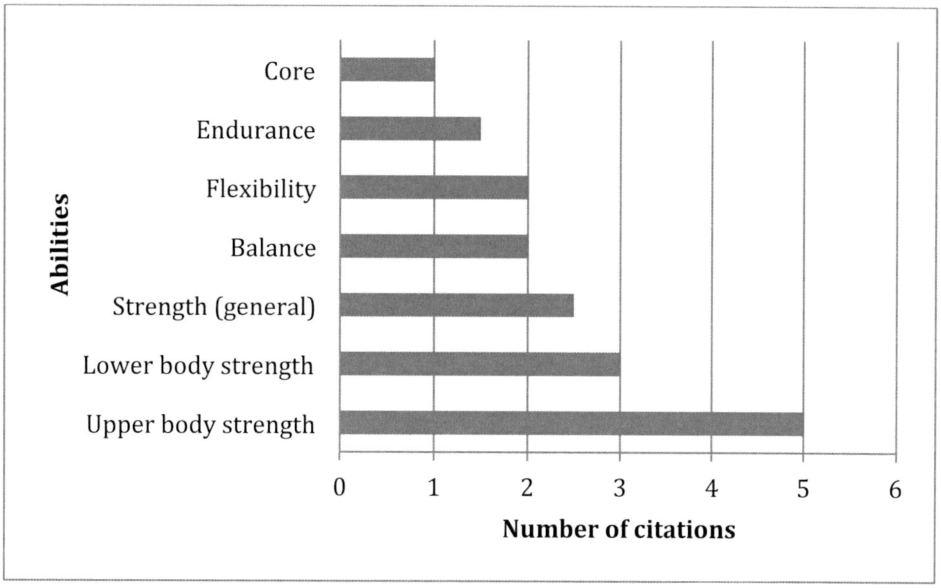

Figure 4.6. Tasks Most Often Cited in PJ Critical Incident Interviews

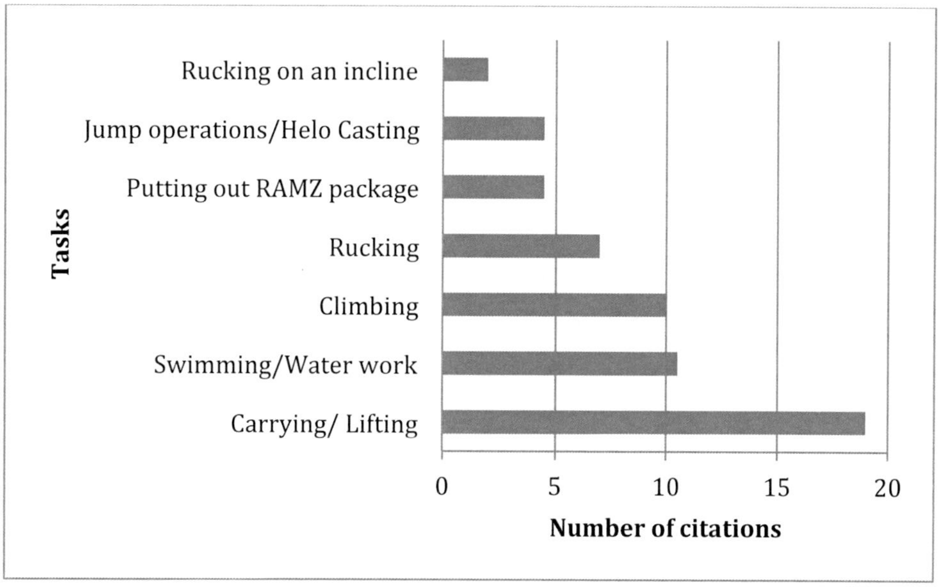

Table 4.4. Sample of Critical Incidents Described During PJ Critical Incident Interviews

Task	Number of citations	Examples
Carrying/lifting	19	• Hauling bodies down a mountain after an helicopter crash • Patient extraction
Swimming/water work	10.5	• Night water work • Dive mission
Climbing	10	• Fast roping • Climbing out of water up a ladder
Rucking	7	• Carrying 70-100lbs of extrication equipment during a ruck march • Walking to a helicopter landing zone
Putting out RAMZ package	4.5	• Picking up operators who did not manage to reach the Rigging, Alternate, Method Zodiac (RAMZ) package • Finning to reach a RAMZ package
Jump operations/helo casting	4.5	• Fast roping off an helicopter • Combat equipment jumps
Rucking on an incline	2	• Overland movement up a mountain • Carrying a litter up and down hilly terrain with ruck on

Movement Pattern Analysis

Five PJ experts, all NCOs, rated the 25 CPTs according to primary movement patterns. The average ratings across all PJ CPTs are depicted in Figure 4.7.

Figure 4.7. Average Movement Pattern Ratings for PJ CPTs

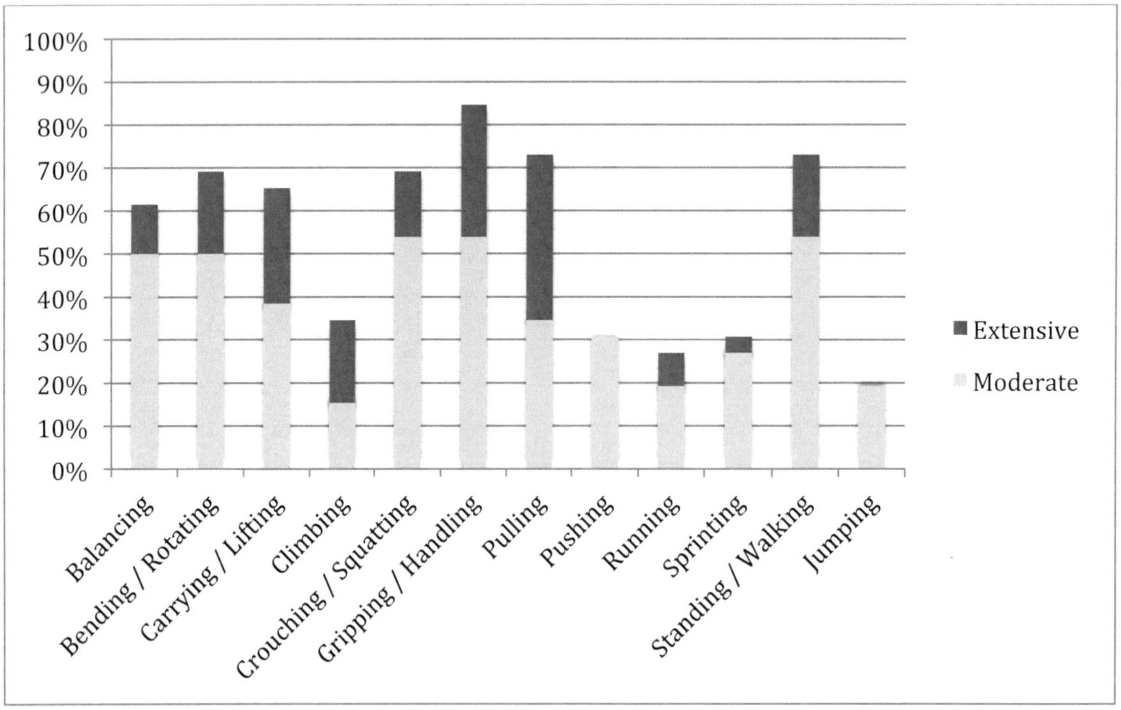

Gripping/handling, pulling, and standing/walking were rated as moderately to extensively used for at least 70 percent of PJ CPTs. Carrying/lifting, crouching/squatting, and bending/rotating were also rated highly. Many of the PJ CPTs that require carrying and pulling involve considerable maneuvering of casualties and equipment to execute rescues. Crouching/squatting and bending/rotating movements are most often required to perform immediate action drills to break enemy contact, to provide medical treatment, and to lift survivors from aircraft or vehicle wreckage. Standing/walking is most often needed to perform land search and reconnaissance activities. The full list of CPTs and average movement pattern ratings are provided in Appendix B.

Physical Ability Analysis

The physical abilities required to perform PJ CPTs, defined by an average SME rating of 3.0 (important) or above on a 5-point scale, are depicted in Figure 4.8. Although all of the abilities are needed to some extent, over 80 percent of PJ CPTs were rated as requiring strength, muscular endurance, and cardiovascular endurance. Due to the extended duration of many of the PJ CPTs, muscular endurance was rated as a required ability for every CPT. Between 40 and 60 percent of CPTs require anaerobic power, equilibrium, and agility. Flexibility was rated as the least important but was still needed to perform 28 percent of the CPTs.

Figure 4.8. Percentage of PJ CPTs Requiring Each Physical Ability

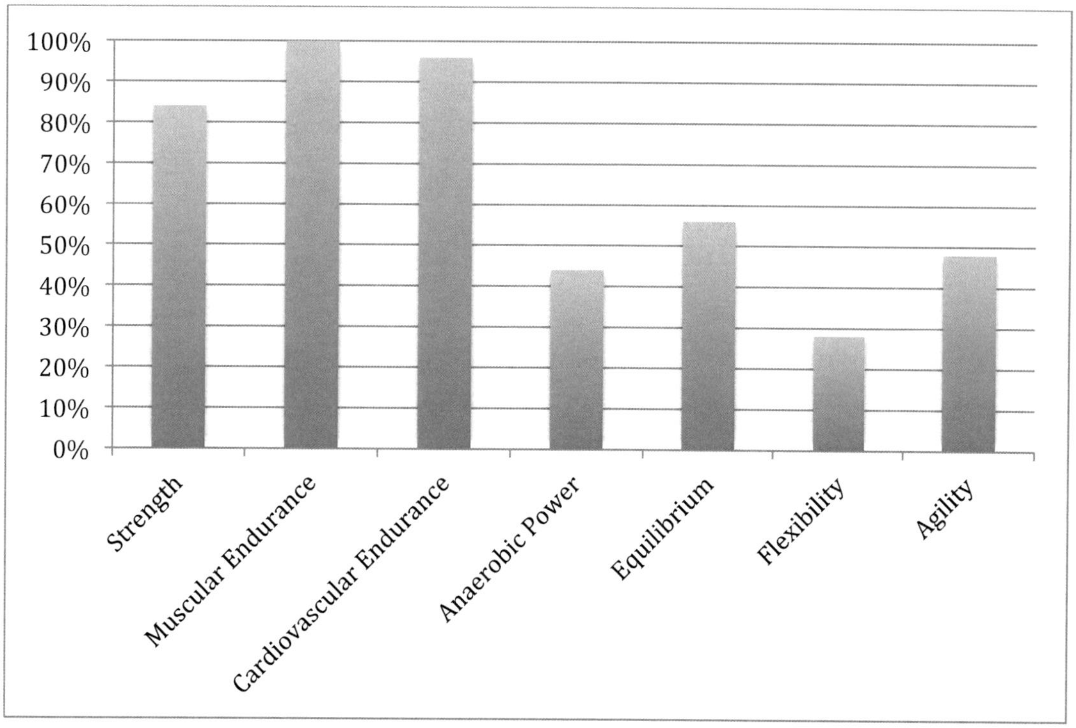

Similar to other Battlefield Airmen, PJs operate and move while carrying heavy loads (e.g., body armor, ruck sack, equipment). These heavy loads require both strength and considerable cardiovascular and muscular endurance to ensure successful performance. The high ratings for these abilities also reflect the movement patterns required for successful performance (as shown in Figure 4.7). Specifically, the strength and endurance ratings are consistent with results indicating moderate to extensive carrying/lifting, pulling, and climbing CPTs. Although flexibility was rated as important for fewer CPTs, this ability is required to carry casualties and to perform specific climbing techniques. For example, PJs may use specialized climbing equipment to maneuver down the side of a steep mountain to reach a casualty.

Special Operations Weather Team

Background

Special Operations Weather Team (SOWT) Airmen provide meteorological, oceanographic, and space environment information while deployed in hostile or denied areas (Air Force Enlisted Classification Directory, 2012). They collect, evaluate, and interpret information from the environment (air, water, terrain) and forecast potential effects on operations. Special reconnaissance and surveillance missions are used to collect some of the environmental data. SOWT Airmen are assigned to Air Force Special Tactics teams or squadrons working with Army Special Operations (U.S. Air Force, 2010a). The largest SOWT squadron, the 10th Combat Weather Squadron (10 CWS), is located at Hurlburt Air Field, Florida.[1]

As special operators, SOWT Airmen are expected to conduct missions in different climates and under various conditions (e.g., day or night, hot or cold, at altitude). These weathermen are trained in infiltration and exfiltration, insertion and extraction, and warfighter tactics, techniques, and procedures (e.g., hand-to-hand combat) (Career Force Education and Training Plan, 2012). These duties require SOWT Airmen execute a variety of physical tasks as part of their work.

Among the four Battlefield Airman specialties examined in this study, SOWT is the newest (established May 5, 2008). However, SOWT Airmen have been a part of the Air Force for many years and were previously described as "combat weathermen."

Focus Groups

Focus groups of SOWT operators were conducted at Hurlburt Field. A total of 16 operators from the 10 CWS participated in four focus groups. Although the sample size is small, the SOWT population is generally small (approximately 75 operators in the Active Component[2]). Moreover,

---

[1] Although active at the time of the study, the 10th CWS no longer exists.

[2] This population consists of assigned five- and seven-level operators in the active component. They make up about 48-percent of the assigned active component personnel in the specialty. Population estimates were based on AFPC personnel extracts from April 2012. SOWT Airmen are also in the Guard and Reserves but in much smaller

a majority of the operators (about 60-percent) are assigned to the 10 CWS. Deployments and training schedules also reduced the number of SOWT operators available for focus groups.

Focus group participants rated 67 tasks according to their level of importance, frequency, duration, and intensity (see Appendix C for full scale). The 67 tasks covered various physical activities, from overland movements to combat jumping and swimming (see Appendix C for the full list of tasks).

Based on participants' ratings of the 67 tasks, we identified 13 CPTs for SOWT operators. These are described in Table 4.5.

Table 4.5. SOWT Critical Physical Tasks (CPTs)

Task	Average Importance
Perform battle drills (e.g., react to indirect fire, evacuate a casualty, react to vehicle roll-over, enter and clear a room)	4.63
Perform live fire immediate action drills (day or night)	4.56
Perform movements under fire (direct or indirect) while mounted	4.56
Perform as a member of a patrol	4.44
Perform immediate action drills (IADs)	4.38
Perform movements under fire (direct or indirect) while dismounted	4.38
Perform reconnaissance or surveillance activities	4.38
Perform small unit tactics (SMUTs) and patrolling	4.27
Move as a member of a fire team	4.25
Perform night movements	4.19
Perform land navigation (day or night)	4.13
Create formations and use defensive fire techniques for overland infiltration and exfiltration operations	4.13
Move through, over, or around obstacles (except minefields)	4.00

NOTE. Importance scale was rated on a scale ranging from 1 (not important) to 5 (crucial). The full scale is presented in Appendix C.

Critical Incident Interviews

We conducted interviews with a total of five SOWT experts, four SNCOs, and one SOWT officer. One was a group interview with two SNCOs and the one officer, although most of the information came from the SNCOs. Figures 4.9 and 4.10 present the most common types of abilities and tasks representing the critical incidents discussed during the interviews. Table 4.6 provides examples of the most common types of physical tasks that were described.

numbers. Based on data from AFPC's Interactive Demographic Analysis System (IDEAS), there were approximately 15 SOWT Guardsmen at the five and seven skill levels in April 2012. According to IDEAS data, there were no assigned SOWT Reservists in April 2012.

Figure 4.9. Abilities Most Often Cited in SOWT Critical Incident Interviews

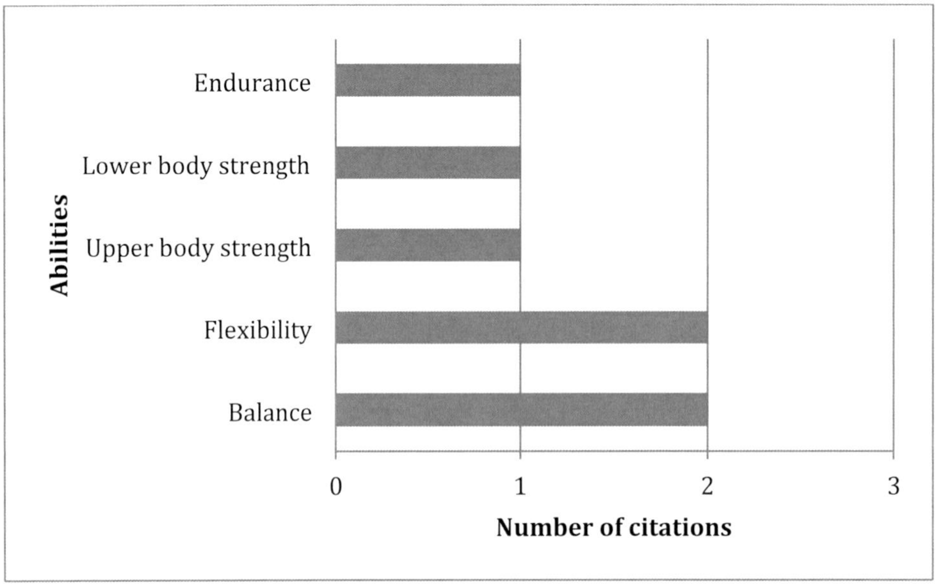

Figure 4.10. Tasks Most Often Cited in SOWT Critical Incident Interviews

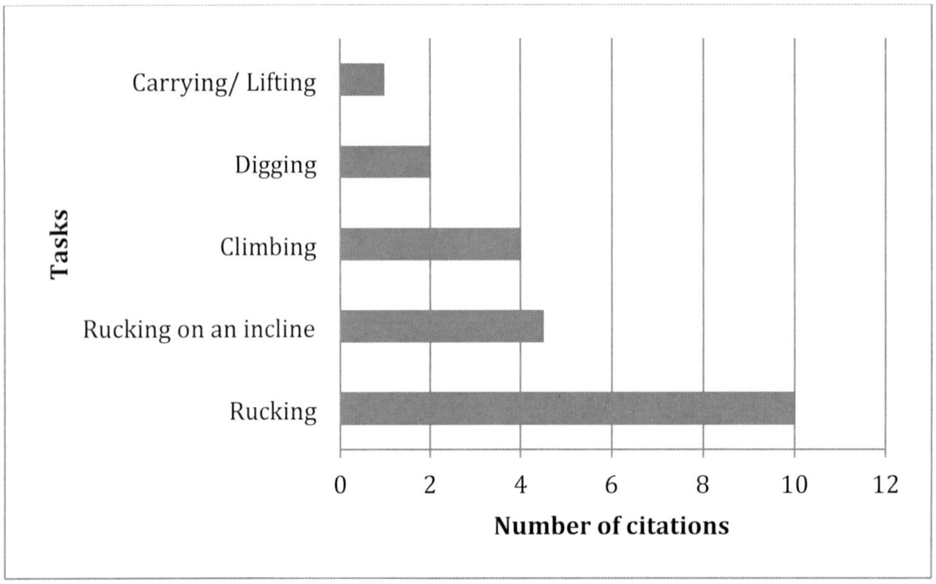

Table 4.6. Sample of Critical Incidents Described During SOWT Critical Incident Interviews

Task	Number of citations	Examples
Rucking	10	• Foot patrols • 18-hour mission of rucking with some sprints
Rucking on an incline	4.5	• Moving up the side of a mountain • Deployment in hilly area
Climbing	4	• Climbing walls • Scaling terraces
Digging	2	• Digging a defensive position
Carrying/lifting	1	• Carrying weather sensors • Carrying heavy weapons or ammunition

Movement Pattern Analysis

Four SOWT experts (two SOWT officers and two SOWT SNCOs) rated the 13 CPTs according to primary movement patterns. The average ratings across all SOWT CPTs are depicted in Figure 4.11.

Figure 4.11. Average Movement Pattern Ratings for SOWT CPTs

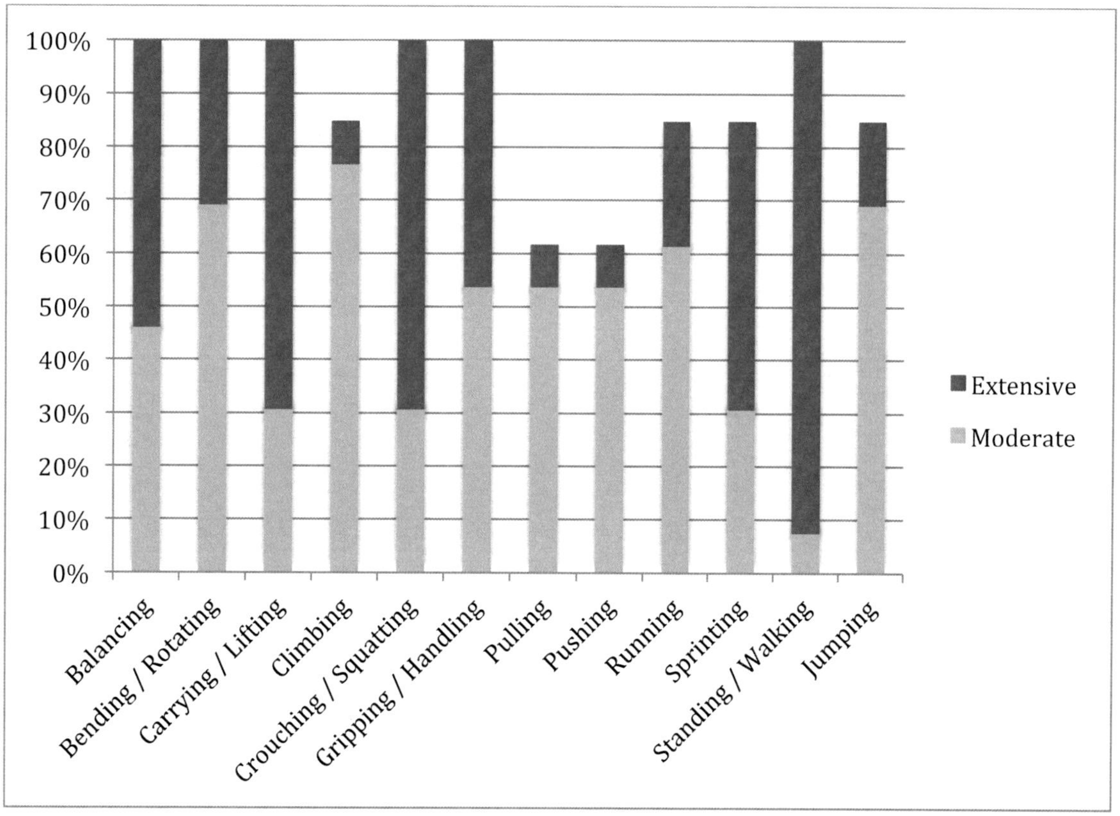

Standing/walking, crouching/squatting, and carrying/lifting were rated as extensively used for at least 70 percent of SOWT CPTs. Balancing and gripping/handling also received high ratings. Many of the SOWT CPTs involved movements by foot while under load (patrols, small unit tactics, overland infiltration), hence the very high ratings for standing/walking and carrying/lifting. Crouching/squatting and balancing are frequently needed to perform tasks such as immediate action drills, battle drills, moving around objects, and reacting to fire while dismounted. Reacting to fire (mounted or dismounted), small unit tactics, creating formations, and battle drills were among the tasks needing extensive use of gripping/handling. Of those tasks, battle drills had the highest rating. Battle drills involve gripping as operators drag casualties from the scene, pull things aside to enter and clear a room, among other things. The full list of CPTs and average movement pattern ratings are provided in Appendix B.

Physical Ability Analysis

The physical abilities required to perform SOWT CPTs are depicted in Figure 4.12. Similar to the other three specialties, over 80 percent of SOWT CPTs were rated as requiring strength, muscular endurance, and cardiovascular endurance. Between 50 and 80 percent of CPTs require equilibrium. The remaining abilities—anaerobic power, flexibility, and agility—are required for only 20 to 50 percent of CPTs.

Figure 4.12. Percentage of SOWT CPTs Requiring Each Physical Ability

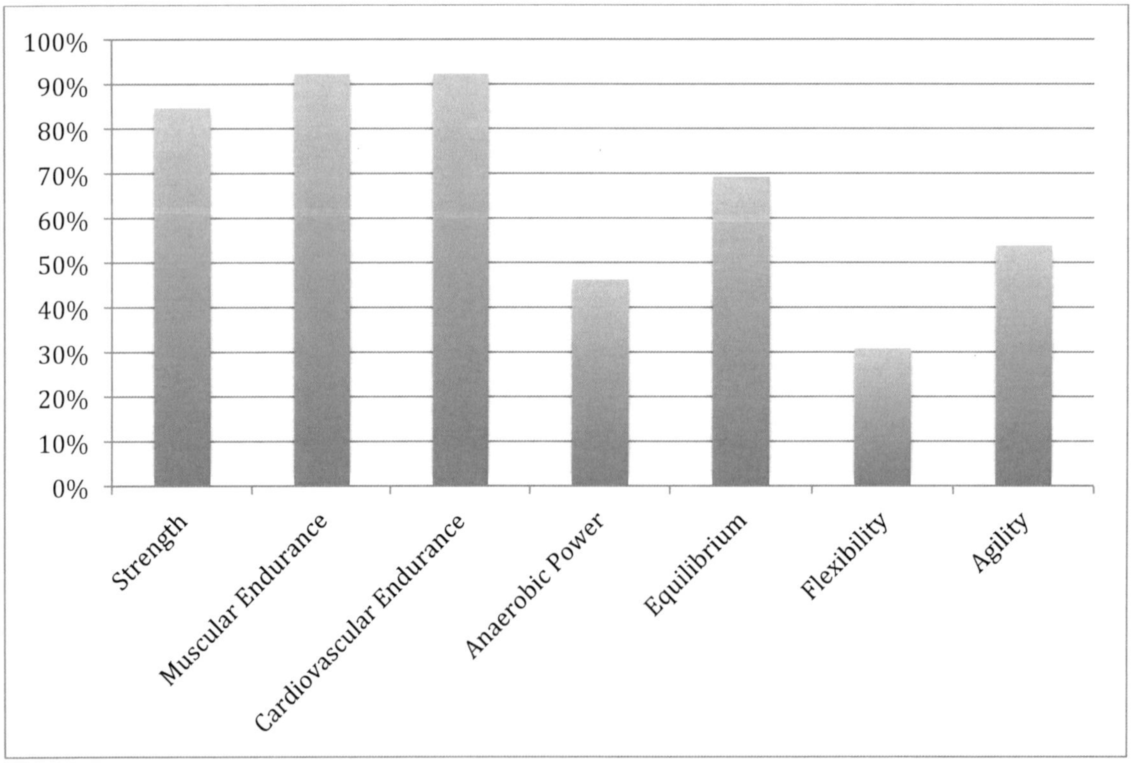

The nature of the SOWT CPTs help explain these findings. Many of the CPTs involve dismounted movements under load, some for significant periods of time and distances. Strength, muscular endurance, and cardiovascular endurance are critical for these types of movements. Equilibrium is needed for tasks that involve balancing, as described in the movement pattern and critical incident analyses (e.g., moving up a mountain). The lower ratings for anaerobic power, flexibility, and agility suggest relatively less use of short bursts of strength (jumping, sprinting), stretching or reaching, and quick movements in different directions.

Tactical Air Control Party

Background

Tactical Air Control Parties (TACPs) advise ground commanders on the use of air power and direct close air support. They also control communications with the air space and control air traffic. They are assigned to conventional Army combat units as well as, in some instances, Special Forces, Navy SEALs, and Army Rangers (Air Force Special Tactics [24 SOW], no date). TACPs are composed of Joint terminal attack controllers (JTACs) who can call for close air support, and recon, observe, mark and destroy (ROMADs), who are training to become JTACs and assist them by driving vehicles, carrying and maintaining equipment, and operating radios.

In addition to setting and operating communication equipment, and calling for air support, TACPs engage in a variety of physical tasks, including conducting infiltration and exfiltration, performing mounted and dismounted operations, engaging in small unit tactics, and evacuating casualties when needed (Department of the Air Force, 2009). They also carry communication equipment weighing up to 30 pounds on missions that can last for several days.

Focus Groups

Focus group participants (i.e., operators) rated a total of 69 tasks according to their level of importance, frequency, duration, and intensity (see Appendix C for full scale). The 69 tasks listed covered a wide range of type of activities, from dismounted operations to insertion/extraction or assault missions (see Appendix C for full list of rated tasks).

The research team conducted focus groups with a total of 114 TACPs from six squadrons (see Table 4.7). The selection of locations aimed at providing the largest possible range of types of units for TACPs, with the assumption that TACPS who have deployed with a mechanized or paratrooper unit (for example) may have performed operationally different tasks and/or rate tasks differently.

Table 4.7. TACP Focus Group Characteristics

Squadron	Location	Focus	Number of Participants
5th ASOS	Fort Lewis-McChord, WA	Stryker brigades	22
9th ASOS[a]	Fort Hood, TX	Mechanized and heavy armor	26
14th ASOS	Fort Bragg/Pope, NC	Airborne	25
19th ASOS	Fort Campbell, KY	Air assault	19
20th ASOS	Fort Drum, NY	Light infantry	15
23rd STS	Hurlburt Field, FL	Special operations	6

[a] This focus group was preceded by a half-day demonstration of typical TACP training.

Based on participants' ratings, these 69 tasks were reduced to a smaller number of CPTs. The resulting 20 CPTs are listed in Table 4.8.

Table 4.8. TACP Critical Physical Tasks (CPTs)

Task	Importance
React to enemy contact	4.85
React to improvised explosive devices	4.83
React to far/near ambush	4.83
React to direct/indirect fire	4.80
Evacuate injured personnel from vehicle	4.69
Navigate by foot during day or night operations using maps or compasses	4.63
Conduct dismounted operations	4.57
Perform foot marches	4.52
Perform small unit tactics	4.47
Break contact with the enemy	4.46
Perform offensive actions	4.44
Conduct infiltration, surface movement, and exfiltration functions with combat maneuver force	4.44
Perform team and squad movement techniques	4.38
Perform vehicle egress under roll-over conditions	4.37
Perform infiltrations or exfiltrations	4.34
Perform defensive actions	4.32
Enter and clear a building	4.25
Perform survival, evasion, resistance, and escape (SERE) activities	4.20
Perform individual movement tactics	4.16
Perform assault zone operations	4.08

NOTE. Importance scale was rated on a scale ranging from 1 (not important) to 5 (crucial). The full scale is presented in Appendix C.

Critical Incident Interviews

Critical incident interviews were conducted with a total of seven TACPs, all of them SNCOs or senior officers. Figures 4.13 and 4.14 present the most common types of abilities and tasks representing the critical incidents discussed during the interviews. Table 4.9 provides examples of the most common types of physical tasks that were described.

Figure 4.13. Abilities Most Often Cited in TACP Critical Incident Interviews

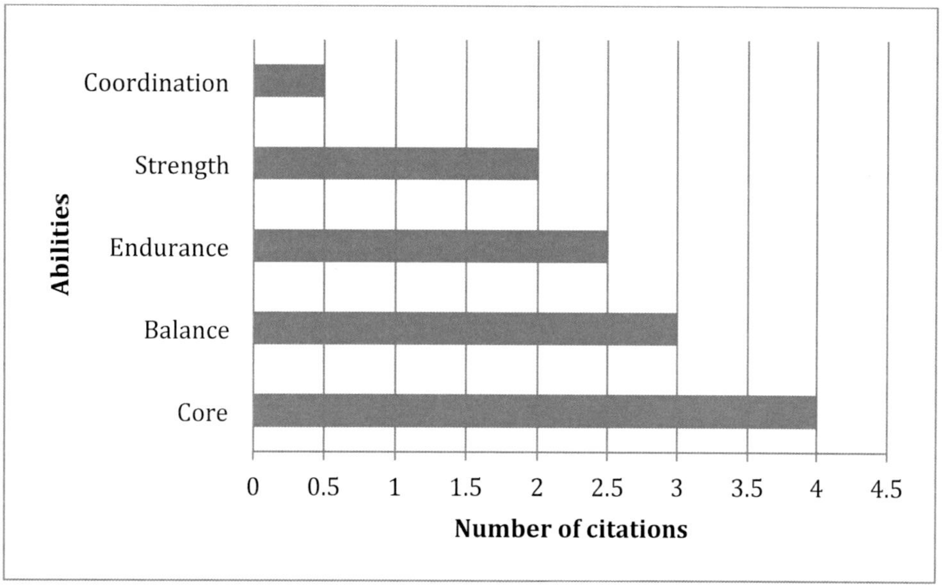

Figure 4.14. Tasks Most Often Cited in TACP Critical Incident Interviews

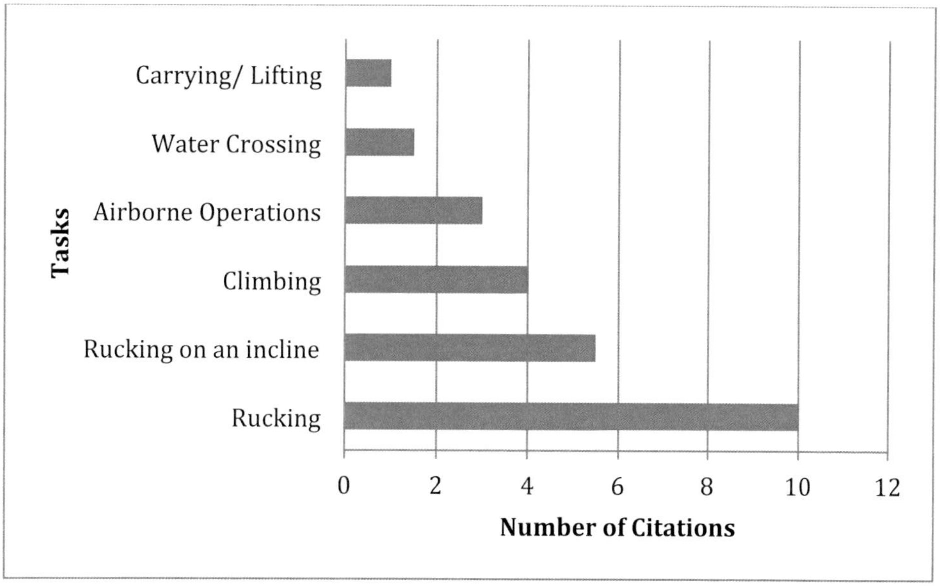

Table 4.9. Sample of Critical Incidents Described During TACP Critical Incident Interviews

Task	Number of citations	Examples
Rucking	10	• Not being able to follow the Army unit's pace while rucking • Mission in Afghanistan where individual cannot carry his load
Rucking on an incline	5.5	• Falling backward off a mountain • Combat mission from 4,000 to 9,000 feet with a 100-lb. load
Climbing	4	• Setting up a tower with an Army communication unit • Climbing stairs
Airborne operations	3	• Airborne operations resulting in broken legs, sprained ankles, twisted knees • Air assault mission where someone has to be evacuated for dehydration because his load is too heavy
Water crossing	1.5	• Crossing 400-500 meters in a river with gear on
Carrying/lifting	1	• Lifting someone out of a burning vehicle

Movement Pattern Analysis

Five TACP experts (NCOs) rated the 21 CPTs according to primary movement patterns. The average ratings across all TACP CPTs are depicted in Figure 4.15.

Figure 4.15. Average Movement Pattern Ratings for TACP CPTs

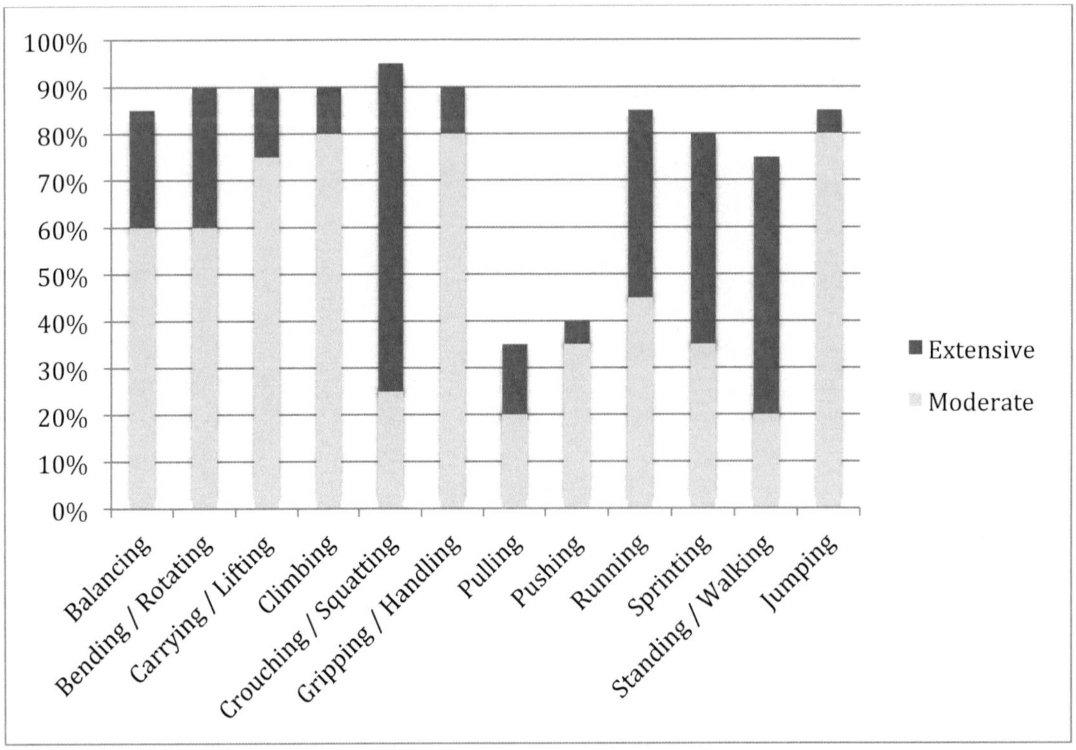

Standing/walking and crouching/squatting were rated as extensively used for at least 50 percent of TACP CPTs. Running and sprinting also received high ratings. These movement patterns were rated as necessary to react to a number of different situations, including reacting to enemy contact, improvised explosive devices, ambush, and indirect fire. Some TACP CPTs required a range of movement patterns. For example, performing individual movement tactics requires balancing, bending/rotating, carrying/lifting, crouching/squatting, and standing/walking. The full list of CPTs and average movement pattern ratings are provided in Appendix B.

Physical Ability Analysis

The physical abilities required to perform TACP CPTs are depicted in Figure 4.16. Similar to the other three specialties, over 80 percent of SOWT CPTs were rated as requiring strength, muscular endurance, and cardiovascular endurance. Eighty percent or more CPTs also require anaerobic power, equilibrium, and agility. Flexibility is needed for only 25 percent of the CPTs. The high number of CPTs that require running and sprinting for reacting to different situations helps to explain the relatively higher emphasis nature on agility and anaerobic power compared with the other Battlefield Airman specialties. Similar to the other specialties, muscular strength and endurance are important abilities for conducting different movements (e.g., rucking) while carrying heavy loads.

Figure 4.16. Percentage of TACP CPTs Requiring Each Physical Ability

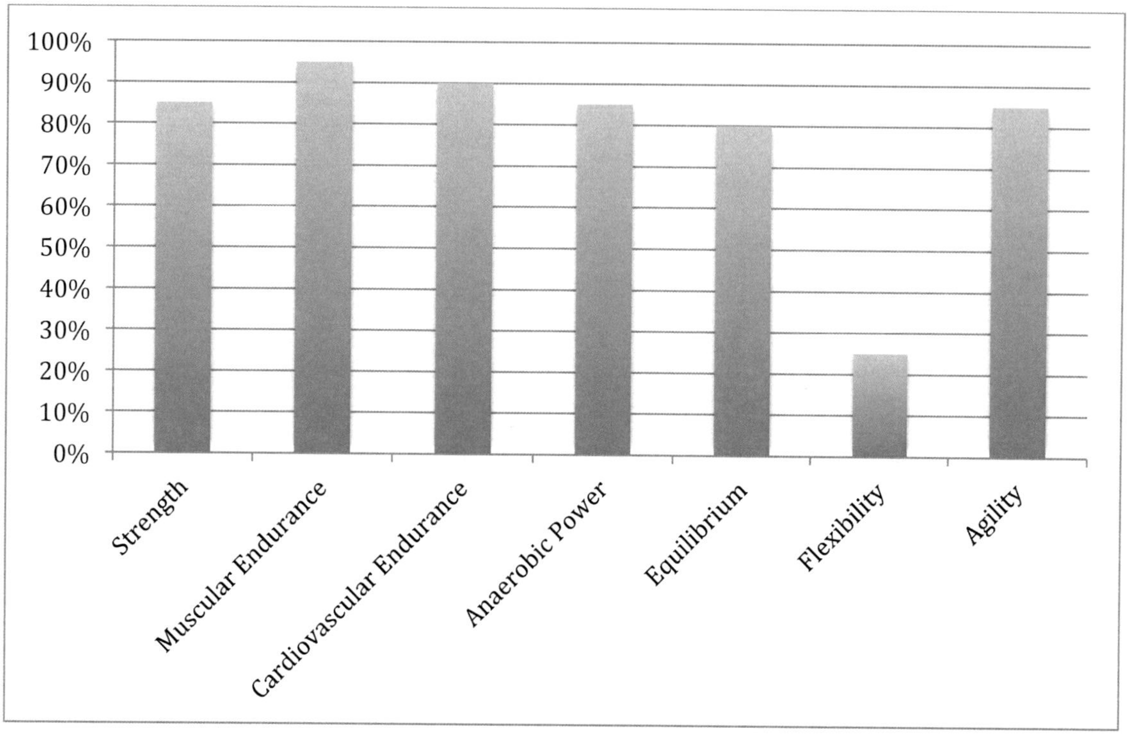

Comparisons by Unit Type and Experience

To address potential concerns about variability in focus group ratings, we planned to conduct a few additional analyses by specialty. Specifically, we wanted to know whether ratings systematically differed by type of unit (i.e., location) and operator experience in the career field (based on self-reported year of graduation from technical training[3]). Lack of evidence showing significant differences by unit type and experience would suggest that the CPT lists can be applied across the specialty. After reviewing sample size requirements for conducting each of these analyses, we were only able to compare the relationship between experience and operator responses for CCTs and TACPs. Much larger sample sizes will be needed to identify potential differences across units within an occupational specialty.

Experience

This analysis involved the relationship between experience (years since technical training) and importance and frequency ratings of CPTs. Specifically, we correlated year of graduation with each type of rating. We analyzed only TACP and CCT ratings because the sample sizes were too small for PJ and SOWT. We had a total of 40 TACP tests (20 CPTs x 2 ratings) and 32 CCT tests (16 CPTs x 2 ratings). For each set of tests, we applied a correction for the family-wise error rate.[4]

Although the TACP correlations were negative or near zero, none were statistically significant. Only one correlation came close: Less experienced TACP operators reported conducting infiltration, surface movement, and exfiltration functions with combat maneuver force more often than more experienced TACP ($r = -0.27$, $p = 0.008$). In general, TACP operators who graduated 2006 or later tended to provide higher frequency ratings than TACP operators who graduated prior to 2005. Although additional data would need to be collected to determine the cause of this relationships, it's possible that less experienced operators were exposed to different mission profiles and deployed operations as the focus shifted from Iraq to Afghanistan.

For CCTs, only one correlation was statistically significant. Compared with more experienced CCT operators, less experienced CCT operators were more likely to rate "Drag casualties from immediate battle area to cover" as important to their jobs ($r = -0.44$, $p = 0.001$). Other correlations for CCT operators were not consistently negative or positive in direction.

In general, there is limited evidence that operator experience related to ratings on the TACP and CCT CPTs. In the cases where there were significant or nearly significant differences, the

[3] Future efforts could also examine alternative measures of experience such as deployed days and training operation days.

[4] For the 40 TACP tests, the correction required p-values less than 0.001 for each test. For the 32 CCT tests, the p-values had to be less than 0.002.

findings suggest that less experienced operators rated CPTs more highly than more experienced operators.

Summary

The physical abilities identified as critical for each of the occupational skills include the following:

- muscular strength
- muscular endurance
- cardiovascular endurance
- anaerobic power (ability to do high-intensity, short duration activity)
- equilibrium
- flexibility
- coordination and agility.

The radar chart in Figure 4.17 shows the relative importance of the physical abilities for the four specialties as determined by focus groups and interviews. Lines closer to the center of the diagram indicate that fewer CPTs in that specialty require a particular physical ability and those closer to the outer edge indicate more tasks. It shows a high demand (more than 80 percent of the tasks) across the four specialities for strength and muscular and cardiovascular endurance, followed by agility, anaeroboic power, and equilibrium. Flexibility was not deemed as critically important, only being required for 20 to 50 percent of CPTs.

Figure 4.17. The Relative Importance of Battlefield Airman Physical Abilities

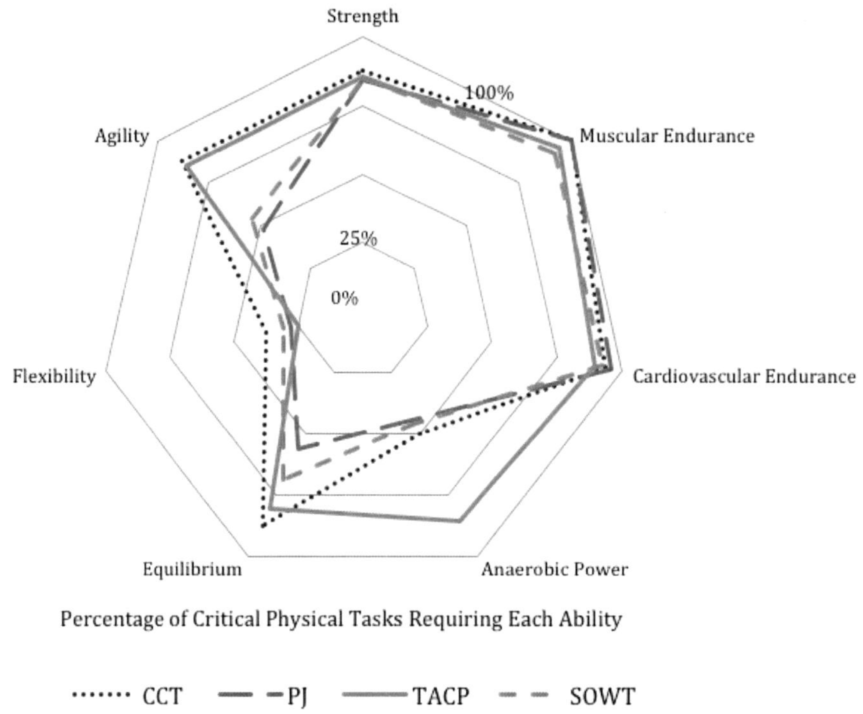

The data indicate that over 80 percent of the CPTs performed by each specialty require muscular strength, muscular endurance, and cardiovascular endurance. Equilibrium was required for at least 50 percent of the CPTs across specialties. Although still required to perform some CPTs, flexibility was only needed to perform between 20 and 50 percent of the CPTs across specialties.

Note that the importance of these abilities assumes equal importance across the CPTs. Although we selected only physically demanding tasks that met a high threshold for importance (i.e., rating of 4 out of 5), it is likely that the importance of CPTs will vary by mission requirements. For example, additional CPTs may emerge as conflicts shift to new environments. Therefore, ratings on the importance of physically demanding tasks should be updated regularly as required by anticipated changes in physical demands.

5. Recommendations for Developing Occupationally Relevant Fitness Tests

This chapter combines results from this study with our literature review to provide recommendations to the Air Force in designing its next steps in conducting validation studies and further identifying physically demanding occupations that may require Tier II fitness standards. As described in Chapter Two, validation studies will provide the data necessary to (a) ensure that tests measure important physical abilities required for successful mission/job performance, (b) ensure that performance on tests are good indicators of mission/job performance, and (c) identify minimum test standards that are associated with acceptable mission/job performance. In addition to meeting these objectives, we also consider required resources, potential injuries, and efficiency in developing recommendations.

In this chapter, we offer an overarching recommendation, followed by three specific recommendations. We conclude by providing final thoughts on how the Air Force can use the findings of this study and moving forward with Tier II standards for other occupational specialties.

Overarching Recommendation: Conduct a Criterion Validation Study to Establish Tier II Standards

In designing a validation study, we recommend starting with a criterion-related strategy to identify the relationship between fitness test scores and job, mission, or training performance on physically demanding tasks. The validation study should be conducted for each occupational specialty following the steps outlined below. In combination with the results identifying the CPTs from this study, a systematic review should be conducted to identify the best options for screening operators on their physical readiness. That is, research should be undertaken to identify potential combinations of physical ability tests and/or work simulations to ensure that all operators are physically prepared to perform their jobs. Particular attention should be given to test reliability, cost and ease of administration and implementation, and, most importantly, coverage of the important physical abilities and tasks performed by operators.

Once tests are selected for the study, the second step is to establish or develop appropriate job and training performance measures. This will require additional work and collaboration with career field managers, squadron commanders, and NCOs (i.e., SMEs). Discussions with squadron commanders and feedback from the Human Performance Summit[1] we attended indicated that operator success is very difficult to define. To address this difficulty, we recommend developing a behaviorally based performance evaluation scales for each Battlefield Airman specialty. A behavioral observation scale (BOS) allows raters who are familiar with operators' performance to identify the frequency that physically demanding tasks are performed. For example, a rater may evaluate how often an operator "keeps up with his team on overland movements" or "carries others team members' gear when they are fatigued." Research indicates that BOSs are reliable and effective methods for measuring performance (Latham and Wexley, 1993).

Once BOSs are developed, the next step is for SMEs to define the minimally acceptable level on the BOS. This will help to identify any potential discrepancies in leader expectations of operator performance. The final steps, four and five, should be accomplished by a trained analyst with a background in personnel selection and assessment.

1. Select a battery of physical ability tests.
2. Identify appropriate measure of job/mission performance.
3. Establish minimally acceptable level of job/mission performance (e.g., reach objective within 15 minutes of target time).
4. Analyze relationship between tests and performance.
5. Establish minimum score(s) on tests.

Use Multiple Tests to Measure Each Physical Ability

To have the most options for selecting an optimal combination of tests, we recommend using at least two tests to measure each ability. These tests may include a combination of basic fitness tests and simulations and can be integrated with the current Air Force Special Operations Command PJ operator tests and with tests conducted by strength and conditioning coaches assigned to the different squadrons. However, it is critical to ensure that any test considered for validation follows a strict protocol to ensure consistent administration, scoring, and reporting.

Although flexibility was identified as an important ability for each Battlefield Airman specialty, we do not expect flexibility to relate strongly to performance. In fact, the quantitative review suggests that flexibility is among the weaker indicators of performance. However, flexibility has been cited as a potential indicator of injury risk; therefore, the Air Force may consider evaluating the potential benefits of a flexibility test (e.g., FMS) by using injuries as the criteria.

[1] A Human Performance Summit was organized by the 23rd Special Tactics Squadron and included both civilian and military subject-matter experts from a wide range of backgrounds to discuss ways to address the limitations of current physical fitness training and testing.

Use Simulations to Offset Body Weight Bias of Basic Fitness Tests

Many basic fitness tests (e.g., pull-ups, three-mile run) are potentially biased in favor of smaller, leaner operators. The extent to which this bias occurs can result in inaccurate decisions regarding the physical readiness of Battlefield Airmen. Therefore, we recommend integrating a job simulation that samples CPTs. Although specific simulations can be developed for each career field, we recommend developing a simulation that includes CPTs that are shared among all Battlefield Airman specialties. For example, CPTs shared across specialties include a ruck march and casualty drags or carries.

An important factor to consider in developing a simulation(s) is to match as closely as possible the physical demands of the operational environment. For example, operators should perform simulation tasks wearing similar clothes, body armor, and ruck sack. Although every mission differs in its physical demands, a simulation should attempt to match what is typical or reasonably expected. Therefore, a simulation should not require operators to ruck the longest distance ever recorded, nor should they be expected to lift or carry the heaviest person in the U.S. military.

Additional guidance and discussions of the challenges in using simulations emphasize the importance of sampling from job tasks, standardizing test conditions, and allowing test takers to practice the test (Callinan and Robertson, 2000; Henderson, Berry, and Matic, 2007; Lievens and De Soete, 2012).

Test Alternative Methods for Identifying Optimal Test Battery

After operators have been scored on the various tests and simulation(s), steps can be taken to identify the optimal combination of tests needed to effectively determine physical readiness. Although a variety of different methods to combine test scores could be considered, we recommend using a compensatory model. A compensatory model allows stronger performance on one or more tests to make up for slightly weaker performance on other tests. Although allowing weaker performance on some tests may seem counterproductive to reach physical readiness goals, a compensatory model may help determine whether operators have a combination of physical abilities that results in effective performance of CPTs.

This type of model is more likely to resemble how military tasks are performed. That is, operators may have very different physical ability strengths and weaknesses, but they can still accomplish the mission safely and effectively. Some operators may have strong upper bodies, whereas other operators may have strong lower bodies. In some cases, one ability (e.g., cardiovascular endurance) may not compensate for another ability (e.g., a lack of muscular strength). In these cases, a true minimum test score would be recommended.

A second advantage of a compensatory model is that an optimal number of tests can be identified. At some point, adding tests provides no additional information on the physical readiness of operators. Going beyond this point increases the cost of testing, the time required for administration, and the potential injury of operators. Finally, a compensatory model can be used to create a single overall physical readiness score by more strongly weighting tests that are better indicators of physical readiness.

Final Thoughts

How Our Findings Can Inform Occupationally Relevant Gender-Neutral Standards

This report outlines the steps and provides an example with four Battlefield Airman specialties of how to identify the job-specific physical demands and the physical abilities needed to perform those tasks. The information provided by these analyses is only the first step toward establishing occupationally relevant physical fitness standards. The next step, as detailed in the recommendations above, is to conduct a validation study. The validation study will provide the necessary information to determine the appropriateness of each physical ability test and how best to combine those tests to determine whether Battlefield Airmen are physically capable of performing their jobs.

Although this study focused on occupations closed to female Airmen at the time, the approach we took to set the foundation for developing occupationally relevant fitness standards and our recommendations for a validation study are relevant to the issue of women entering previously closed occupations. Recent changes in DoD policy excluding women from certain assignments and specialties add urgency to the need for the services, including the Air Force, to establish appropriate gender-neutral standards for military occupations. Setting fitness standards that are tied to physical job performance is a key element to setting gender-neutral standards. Our study and recommendations can therefore inform efforts that the Air Force can take to address changes to the combat exclusion policy.

Tier II Standards for Other Occupational Specialties

Not all Air Force specialties that have physical demands will benefit from having additional Tier II standards. Specific criteria should be developed to identify which occupations are in need of additional annual or semiannual standards. Specific criteria could evaluate how much physical effort is required to perform job tasks, whether these tasks are essential to the job, and whether tasks can be easily modified using equipment or the assistance of other personnel. Additional guidance and feedback from squadron commanders could also identify whether their Airmen are having difficulty performing physical tasks or whether Airmen are getting injured on the job as a result of their poor fitness.

Furthermore, it is likely that only a small subset of physically demanding specialties need to use regular Tier II fitness tests to ensure physical readiness. As previously discussed, other methods are available to identify Airmen who are not physically capable of performing their jobs. If the majority of the physically demanding work is repetitive (e.g., loading cargo), supervisors are generally a more efficient and timely source for identifying underperforming Airmen. In contrast, Battlefield Airmen may only occasionally perform some of their required CPTs. For example, TACPs do not have to "break contact with the enemy" on a regular basis but still need to maintain the physical abilities to perform this task successfully when it occurs. A similar example is provided with police officers who may never shoot their weapons but still need to maintain this very important capability. Consequently, the Air Force might consider using its limited resources to develop Tier II tests and standards for those occupational specialties that perform CPTs irregularly.

Appendix A. How to Identify Physical Requirements

Objectives of a Job Analysis

The purpose of this appendix is to highlight what job analysts have available and what types of factors they have to consider when selecting the tools they can use. The first step in determining the physical demands of an occupation is to conduct a job analysis, a systematic process of gathering, evaluating, and making inferences about work activities, worker attributes, and the work context. The results of a job analysis can provide a foundation of evidence for making informed personnel decisions, such as what worker attributes are desired in job candidates. This appendix reviews approaches to job analysis for physically demanding occupations. We specifically focus on approaches from three scientific disciplines: industrial-organizational psychology, biomechanics (including its application to ergonomics), and physiology. We conclude the appendix with an assessment of four data collection methods.

Approaches to Analyzing Physical Job Demands

The three scientific disciplines have typically approached analysis of physical job demands in different ways (Table A.1). Biomechanical approaches are typically used with the purpose of preventing job injuries or musculoskeletal disorders, such as lower back disorder. Underlying this purpose is the load-tolerance model, which states that a load (i.e., force) that exceeds the body's ability to handle the load (i.e., body's tolerance) can result in injury or illness (Marras, 2003). Thus, the central measures for biomechanical analysis are forces, including those external to the body (e.g., force created by lifting a box) and inside the body (i.e., force produced by muscles). Other common measures focus on factors that affect the sizes of forces being exerted, such as body posture, muscular strength, and task duration. Biomechanical approaches have been widely used for tasks involving manual materials handling because they are associated with various musculoskeletal problems, such as lower back pain (Bernard, 1997). These approaches are also popular among ergonomists, who study human interactions with physical work environments (e.g., equipment). Ergonomists use results from a biomechanical analysis to redesign the task or workspace to prevent or reduce physical strain.

Industrial-organizational psychological approaches are usually conducted for setting job selection or classification standards, with the criterion being successful performance on the job (or, in some cases, successful performance in job training or military missions). The central measures for these approaches are ratings of job tasks (e.g., frequency of performing tasks) or worker characteristics (namely, physical abilities) that are tied to successful performance. Many job analysis procedures used in this field include reviewing job or training documents to identify tasks and duties, interviewing SMEs, observing incumbents performing their jobs, and constructing, administering, and analyzing results from job analysis questionnaires. Moreover, industrial and organizational psychologists have a long history of using statistical techniques, such as factor analysis, to classify the ability requirements of physically demanding jobs and occupations (e.g., Fleishman, 1964).

Finally, physiological approaches to analysis of physically demanding jobs focus on the physiological demands or costs of doing the tasks. The purpose of this approach is to measure work capacity—the capability of an individual body to perform a particular physical activity (e.g., run 50 meters)—and reduce physical strain associated with job tasks. Physiologists use a variety of physiological measures to measure demands placed on the cardiovascular and respiratory systems, such as heart rate and oxygen uptake. Physiologists also measure other physiological responses to physical activity, such as hormonal responses and core body temperature. Environmental factors that affect work capacity, such as ambient temperature, altitude, and humidity, are also commonly measured in a physiological analysis of physical job demands. Because of its focus on work capacity, a physiological approach to analyzing physical job demands is particularly well suited for quantifying cardiovascular and muscular endurance. Table A.1 summarizes the objective and common measures of each approach.

Table A.1. Approaches to Analyzing Physical Job Demands

Approach	Primary Objective	Common Measures
Biomechanical/ Ergonomic	Injury prevention	• Forces exerted against and within the body • Factors affecting magnitude of force
Industrial-organizational psychology	Successful performance • Job • Training • Mission	• Ratings of job duties, tasks • Physical abilities linked to performance
Physiological	Work capacity Reduce physical strain	• Physiological demands (e.g., heart rate, oxygen uptake) • Environmental factors associated with reduced work capacity (e.g., humidity)

Analysts of physical job demands have blended methods and objectives from the different approaches. For example, a biomechanical analysis might begin with a task analysis, a popular method in the industrial-organizational psychology approach. Task analysis identifies and describes the job activities required to perform the job (e.g., "install truck engine") and often involves incumbent ratings of task features (e.g., frequency, importance, duration, and difficulty). Likewise, an industrial-organizational psychologist might employ physiological assessments during job analysis of physically demanding jobs. The objectives of the job analysis generally dictate which combinations of methods are used by job analysts. For example, task analysis is helpful if one of the objectives is to understand the content of the job, training, or mission. If injury prevention is a focus, biomechanical and physiological assessments are often used. Indeed, Rayson (2000) recommends that a job analysis of physically demanding occupations, or what he calls a Physical Demands Analysis (PDA), uses a combination of approaches. Specifically, Rayson recommends starting with a task and task-element analysis[1] using observations, questionnaires, interviews, or other methods and then quantifying the physical demands of the tasks with physiological, biomechanical, and psychophysical rating methods.[2]

Collecting Data on Physical Job Demands

Analysis of physically demanding jobs often includes characteristics of the tasks, objects handled during task completion, and the work environment. Task characteristics include type of movement (e.g., lifting), frequency, intensity, duration, body posture, and the objects and equipment involved. Characteristics of the objects include dimensions (e.g., size of surface area), weight, coupling (how it is grasped), and stability. These are of considerable importance in biomechanical analyses. Environmental characteristics include ambient temperature, humidity, noise, altitude, air pollution, workspace, and clothing. To collect information on any of these job characteristics, analysts can use a variety of approaches.

The following sections highlight some of the measures or methods used as part of job analysis of physically demanding occupations. We start with biomechanical and ergonomic methods, followed by industrial-organizational psychology methods and physiological methods. We end with psychophysical measures because they combine biomechanical, psychological, and physiological elements.

Biomechanical Approach

Body Posture

Methods for measuring body posture vary in their levels of biometric accuracy and sophistication. On the lower end of the continuum are direct observational methods, such as the Ovako Working Posture Analysis System or OWAS (Karhu, Kansi, and Kuorinka, 1977). The observer (or a video camera) records the worker's postures for the back, arms, and legs, as well as the load handled or force exerted during the activity. The recordings are converted to four-digit codes used for analyzing combinations of postures. Results from the analysis can reveal patterns of poor working postures used during different types of work activities. The OWAS method has been successfully applied to jobs in different industries and exhibits high inter-

[1] A task element is an underlying movement associated with a task. For example, the task of installing a truck engine involves several elements, such as lowering, holding, turning, and pushing.

[2] Psychophysical rating methods assume biomechanical and physiological stresses on the body while completing physical tasks can be accurately assessed using subjective ratings of perceived stress or effort.

observer reliability (Mattila and Vilkki, 2003). However, the method is best for jobs with easily observable postures. McAtamney and Corlett (1993) created a similar procedure for examining risk of upper limb disorder called the Rapid Upper Limb Assessment (RULA).

Biometric accuracy can be increased by quantifying body segment angles during task performance. Photographic or video images from different angles can be used to get a two- or three-dimensional view of the angle at different body joints (e.g., knees). However, combining images from different angles requires caution because of parallax error, or the illusion that an object has a different position than it assumes in reality because different lines of sight are used to view it. Even greater accuracy can be gained by using goniometers and torsiometers, which are instruments used to (mechanically or electronically) measure angular motions. Some electronic goniometers can be applied to joints (e.g., wrists) without hindering use of the joint, making them well suited for studies in work settings.

Motion capture is the most sophisticated method of measuring body posture. Multiple cameras track movements of participants wearing body suits with markers placed at anatomical areas (Quesada, 2003). Motion capture provides dynamic information about changes in body posture but analysis of the data is a "long and tedious process" (Rayson, 2000, p. 88). Moreover, suits with markers and multiple cameras can be intrusive in a normal work environment.

Force Measurement

To measure forces, biomechanical experts take several measurements of the objects being handled, the position and movement of those objects relative to body joints, and muscle involvement. Object characteristics include its dimensions (e.g., size of surface area), weight, coupling (how it is grasped), and stability. The simplest way to measure loads is to weigh the objects being handled. In the case of pushing and pulling activities, a force transducer is placed between the worker and the object being pushed or pulled. Movement speeds and acceleration need to be controlled or at least accounted for when measuring force exertion.

To measure forces on the body and moments of joints, biomechanical experts often use simulations of the relevant job tasks. For example, if a job task involves lifting boxes from the ground and placing them onto the bed of a truck, a simulation might involve lifting boxes from the ground onto a shelf that is at the same height as the truck bed. A simulation allows the experts to manipulate the task, environmental, and load characteristics in systematic ways to examine the effect of the changes on the forces exerted on the body. Systematically increasing the weight of the box is one method of varying a box-lifting simulation. See Ayoub et al. (1987) for an example of simulation methods applied to Air Force specialties.

Muscle involvement is directly related to the internal forces that are exerted when a person does a physical task. A common method for measuring muscle involvement has been electromyography (EMG), which measures a muscle's electrical activity by means of electrodes placed on the skin over the main part of the muscle (Rayson, 2000). EMG can also be used to gauge muscle fatigue. Despite its popularity, EMG does not always provide reliable results between individuals and electrode settings and requires extensive expertise to use (Gobel, 2005).

Industrial-Organizational Psychology Approach

Task Analysis Surveys

Surveys or questionnaires are a popular way to obtain ratings of physical tasks associated with successful performance. Surveys are either existing measures that have been validated on several jobs in different industries or developed "from scratch" for the particular jobs or occupations being analyzed. In the latter case, job analysts create surveys from task or task element lists they have developed. The task lists are based on results from various data collection and analysis techniques, including reviews of work documents (e.g., training manuals); interviews or focus groups with job incumbents, supervisors, or other SMEs; and observations of workers performing the job. One such method is the Task Inventory with the Comprehensive Occupational Data Analysis Program (TI/CODAP; Christal, 1974), which was developed by Christal at the now-closed Air Force Human Resources Laboratory. The "TI" part of the procedure uses a two-part questionnaire to obtain job-relevant background information and job-task information (e.g., time spent on task) from job incumbents and supervisors. The "CODAP" part is the computer program that contains the data and enables analysts to manipulate the data from the questionnaire. The Air Force's Occupational Analysis Division (OAD) uses a system akin to TI/CODAP for its occupational analyses of enlisted Air Force specialties (AFS).

Existing survey-based measures that can be useful for describing physical job tasks are Functional Job Analysis (FJA; Fine and Wiley, 1971) and Hierarchical Task Analysis (HTA; Annett et al., 1971). FJA was developed by analysts in the U.S. Employment Service of the Department of Labor to classify jobs in the Dictionary of Occupational Titles (DOT). The most-recent version of FJA includes seven scales to assess what workers do using standardized language: (1) things, (2) data, (3) people, (4) worker instructions, (5) reasoning, (6) math, and (7) language. The scales are anchored with specific behavioral statements and tasks. HTA breaks down tasks into subtasks that are arranged hierarchically. For example, using an automated teller machine (ATM) includes tasks such as preparing the transaction and completing the transaction and subtasks such as entering one's PIN code and removing money from the ATM (Hollnagel, 2006).

Worker and Trait-Oriented Surveys

Some job analysis surveys focus on characteristics of the worker instead of the work tasks. In physical jobs, the focus has largely been on physical abilities required to perform the job successfully. The physical ability requirement method is largely rooted in the work of Fleishman and his colleagues (Fleishman, 1958, 1964; Fleishman and Hogan, 1978; Fleishman and

Quaintance, 1984; Theologus and Fleishman, 1973; Theologus, Romanshko, and Fleishman, 1973). Their work began with factor analytic studies to develop a taxonomy of human abilities needed for successful job performance in a variety of occupations. Within that taxonomy were nine physical abilities, which include four types of muscular strength, two types of flexibility, and three constructs related to gross body coordination and stamina. Based on their taxonomic work, Fleishman and colleagues conducted other studies to develop a survey measure of the abilities, now called the Fleishman Job Analysis Survey (F-JAS; Management Research Institute, 2009).[3] The survey requires SMEs (e.g., job incumbents) to rate the level of each ability needed to perform the job. The ability is defined and contrasted to related abilities (e.g., static versus dynamic strength), and behavioral examples are used to anchor the rating scale. For example, the "static strength" rating scale has the following behavioral descriptions at the low and high ends of the scale, respectively: "Requires use of little muscle force to lift, carry, push, or pull a light object" and "Requires use of all the muscle force possible to lift, carry, push, or pull a very heavy object" (Fleishman and Mumford, 1988, p. 923). The ability requirements scales are used to rate the abilities required to perform on the job successfully and, as such, are not tied to specific job tasks.

Two additional worker- or trait-oriented surveys relevant to analyzing physically demanding jobs are Threshold Traits Analysis (Lopez, 1986) and the Minnesota Job Requirements Survey (MJRQ; Desmond and Weiss, 1973; 1975). Threshold Traits Analysis assesses the relative importance of 33 worker characteristics that are grouped into three ability (physical, mental, learned) and two attitudinal (motivational and social) areas. For each trait, a definition is supplied and survey respondents are asked about the level of complexity it requires, how important it is as a job requirement, and how much influence it has on performance (Algera and Grueter, 1998). The MJRQ is a short, trait-oriented survey that asks SMEs to rate the importance of different work actions or activities for their jobs. The work actions and activities are tied to specific traits, some of which involve physical abilities (e.g., "precise movement of fingers in the handling of very small objects") (Algera and Grueter, 1998, p. 150).

Critical Incidents Technique and Repertory Grid

The critical incidents technique (Flanagan, 1954) and Repertory Grid (Kelly, 1955) use interviews or focus groups to collect job information. For the critical incidents technique, SMEs (e.g., supervisors) describe work scenarios that demonstrated particularly effective or ineffective performance. Each description includes information about the incident's background, what was done by the worker that was particularly effective or ineffective, and the immediate outcome of the worker's behavior. Job analysts categorize the incidents to create dimensions of successful (or unsuccessful) job performance, which can be used to develop performance-criterion measures

---

[3] The scale was previously called the Manual of Ability Requirements Scales (MARS; Fleishman, 1975; Fleishman and Quaintance, 1984).

for validation studies. The Repertory Grid is similar in that it tries to elicit information about effective and ineffective performance. However, the Repertory Grid focuses on the worker in that the SMEs are asked to describe characteristics that differentiate an effective from an ineffective worker. Information from both methods can be used to help identify physical abilities related to successful (or unsuccessful) job performance.

Job Analysis Linkage Model

A job analysis linkage model can be used to tie specific physical abilities to relevant tasks or duty categories. The model has been used to analyze physical demands of firefighter work (Hogan, 1991, p. 790). Developed by Zedeck et al. (1988), this procedure involves the basic steps needed for developing selection tests: identification of all job tasks and the physical abilities and skills needed to successfully perform them, removal of tasks and abilities that are unimportant or not needed on the first day of the job, and linkage of important and necessary abilities and skills to their relevant tasks or task categories. The procedure uses different collection methods: observations, interviews, questionnaires, and panel discussion with job incumbents, supervisors, and job analysts. The result of this job analysis procedure is a collection of critical job tasks and abilities that can be used to develop tests for job selection.

Physiology Approach

Oxygen Uptake

Oxygen uptake (VO_2) is an important measure in a physiological analysis of physical job demands. Rate of oxygen uptake indirectly measures the energy used by the individual during physical activity and is thus a measure of cardiovascular stress. Maximal oxygen update (VO_2max) is an individual's maximal aerobic (work) capacity and is an indicator of physical fitness. Individuals vary in their work capacity. Inter-individual variation increases with task complexity. This means that more individuals are needed to analyze a work group's average oxygen uptake for jobs with complex physical tasks versus simpler physical tasks.

A common way to directly measure oxygen uptake is with gas analyzers. In laboratory settings, a gas analyzer is used while study participants do physical activities that simulate their work tasks. The gas analyzer consists of a face mask, a tube to transport the expelled air, and a machine that collects and analyzes oxygen uptake. Often, gas analyzers in laboratory settings are tethered to electrical or other cords that make them nonportable. In field settings, portable gas analyzers can be used to allow participants to engage in actual work tasks. Portable gas analyzers have smaller analyzer devices that participants carry or strap to their bodies. Despite the benefit of measuring oxygen uptake for actual work tasks, portable gas analyzers can be expensive, require expertise to set up, and cannot withstand intense physical activity (Rayson, 2000).

Heart Rate

Another common physiological measure is heart rate (HR), a measure of cardiovascular (and psychophysiological) strain. It is sensitive to psychological changes (e.g., increased psychological stress), environmental conditions (e.g., warm temperatures), as well as the actual physical demands of the task. HR can substitute for measuring VO_2 when measuring VO_2 is not feasible. However, the relationship between HR and VO_2 for the individuals under study needs to be known to do so. A simulated activity in a laboratory setting can be used to measure this relationship before analyzing the tasks in a field setting, where HR would be taken.

Portable heart rate monitors can be used to assess HR in field settings. Environmental conditions, such as ambient temperature and humidity, are also measured during the activities to improve interpretation of the results. The proportion of maximum heart rate can be estimated and used as an indicator of cardiovascular strain from performing the tasks. If assessing HR during the activity is not feasible, recovery HR can be measured instead. Recovery HR is taken immediately after completion of the activity. Recovery HR is a limited measure because it not only depends on the strain of the activity but also on the fitness of the individual, with fitter individuals recovering faster. However, recovery HR can be used to help establish minimum acceptable standards of aerobic fitness (Rayson, 2000).

Body and Ambient Temperature

Cold, heat, humidity, wind, and other work conditions can stress the body, reducing work performance (Astrand et al., 2003). Therefore, physiologists measure both body and ambient temperatures if there is an expectation of cold or hot work conditions. Core (deep) body temperature can be measured orally, aurally, in the throat, or rectally. A more recent advancement has been the use of pills that individuals swallow and which transmit radio signals of body temperature to an external receiver. The pills pass through the digestive system, so they need not be recovered (Rayson, 2000). Peripheral (skin) temperature is taken with skin thermistors placed at specific locations on the body (e.g., inner thigh). A weighted average of peripheral and core temperatures are taken because of temperature varies for different body tissues.

Ambient temperature and humidity are usually measured as wet bulb globe temperature (WBGT), which is a composite of air temperature (dry bulb), radiant temperature (globe), and relative humidity (wet bulb). If conditions are fairly dry, a wet globe thermometer (WGT) or Botsball thermometer can be used. A Botsball thermometer combines air temperature, humidity, wind, and thermal radiation into a single reading. The thermometer can be attached to a helmet worn by the worker so readings can be taken wherever the work is being conducted. However, brisk walking can alter the readings. Astrand et al. (2003) recommend using the Botsball thermometer to initially survey ambient conditions but then leave the thermometer stationary in a work area with the most representative or extreme temperatures.

Stress Hormones

Psychophysiological strain can also be measured by stress hormones excreted during the work activities. Blood cortisol levels can be measured, but blood tests are limited because they are invasive, expensive, require expertise to perform, and yield results that are difficult to interpret without proper baseline data. Measuring cortisol in saliva has the benefit of being less invasive but still is limited in other ways (i.e., expense, expertise, and interpretability of results). Another method of measuring stress hormone levels is by taking urine samples. Catecholamines, a family of hormones (e.g., adrenaline), can be measured in urine samples. However, urine samples can be drawn only about every two hours, and there are large individual differences in elimination of catecholamines in urine. Astrand et al. (2003) recommend using heart rate information instead as an indicator of psychophysiological strain.

Psychophysical Rating Methods

Psychophysical measures rely on subjective ratings of perceived effort or stress associated with physical tasks. The most popular psychophysical measure is Borg's Rating of Perceived Effort (RPE), or the "Borg Scale" (Borg, 1998). The original rating scale spans from 6 (No exertion) to 20 (Maximal exertion), with each rating "approximately corresponding to heart rate divided by 10" (Rayson, 2000, p. 91). A newer scale is based on a 10-point category ratio scale (CR-10) but is less popular than the original scale. Individuals engaging in a physical activity complete the scale, either at fixed time intervals or at particular target points. The Borg Scale can be used instead of measuring heart rate if doing so is not feasible (Rayson, 2000).

Another psychophysical measure is the Index of Perceived Effort (IPE; Hogan and Fleishman, 1979). The IPE includes 7-point rating scales for each task using either generic anchors or task anchors. Hogan and colleagues have validated the IPE against metabolic and ergonomic work cost criteria (Hogan and Fleishman, 1979; Hogan et al., 1980). Job analysts or supervisors evaluate individual job tasks using the IPE. The results can be used to identify "critical" tasks with physical demands.

Finally, the Body Map (Wilson and Corlett, 1995) allows individuals to identify body regions where they are experiencing discomfort after completing a physical task. Individuals are given two-dimensional outlines of the front and rear of an adult human body. The outlines are sectioned off into regions to make it easier for individuals to identify where they experience discomfort. For each region, individuals can rate the severity of discomfort using a 5-point scale. Rayson (2000) recommends against averaging discomfort ratings by body region across individuals without taking into account individual differences in body size, age, gender, or other factors that could affect discomfort ratings.

Advantages and Disadvantages of Data Collection Methods

The job analyst's "tool kit" has increased in complexity as more has been learned about relative strengths and weaknesses of different methods in different job contexts. In Table A.2, we highlight four categories of methods that are commonly used to collect data on physical job demands and briefly list their strengths and weaknesses.

Table A.2. Primary Strengths and Weaknesses of Four Common Job Analytic Methods

Method	Primary Strengths	Primary Weaknesses
Interviews with SMEs	Provides *detailed* job information from those who know the job well (SMEs)	• Can be time intensive • Prone to small samples, which limits generalizability of results • Information is second-hand
Job surveys	More standardized and efficient than interviews Typically offers more generalizable results than interviews	• Low response rates and response bias can be concerns • Does not provide as much detail as interviews or observations • Information is second-hand
Observations of job (or training) tasks	Offers job analysts first-hand knowledge of how job is done If using physiological or biomechanical tools, analysts can directly measure physical demands	• Resource intensive, especially if using physiological or biomechanical assessments • Logistical issues are common (e.g., weather disruptions, scheduling changes, tasks occurring too infrequently, etc.) • Can disrupt normal work patterns
Lab or field simulations	Analysts can measure physiological or biomechanical job demands in (semi-) controlled setting	• Resource intensive (requires lab equipment and additional expertise to use equipment, participants to go to specific sites for assessments, etc.) • Requires advanced knowledge of job tasks from other methods

The first method, interviews, includes one-on-one and group interviews with job incumbents, supervisors, or other experts on the job being analyzed. The primary strength of interviews is the level of detailed information the analyst can get from individuals who do or have done the job. However, interviews can be time-consuming, typically requiring at least one hour. Focus groups provide efficiencies but still require experts to take time out of their schedules. Because of this, job analysts rarely get a large and representative sample of the job incumbents or supervisors. This limits the generalizability of the results from the interviews.

Surveys, however, can be used to counteract some of the limitations of interviews. If designed well, surveys will provide more standardized job information and for a larger, more-representative sample than is typically achievable using interviews alone. Interviews require phone or in-person contact but surveys can be completed online or by mail, requiring little of the job analyst's time except for survey development and analysis of the results. However, surveys often do not provide the level of detail of interviews and participants can ignore surveys more easily, resulting in lower response rates than is desired.

Both interviews and surveys provide "second-hand" information. That is, the analyst is relying on SMEs to describe accurately what is required for the job. Observations, on the other hand, offer the job analyst a first-hand look of what individuals do on the job, removing the bias of second-hand information from the analysis. Observations can also be coupled with physiological or biomechanical assessments. Thus, the analyst gets direct measurements of physical demands (e.g., VO_2max) while individuals are performing their jobs. Direct measures of physiological and biomechanical performance can be especially useful for training design and developing ergonomic interventions to reduce the risk of injuries on the job. However, these types of assessments can disrupt normal work patterns if they require individuals to wear devices to take measurements. Also, observations are prone to logistical difficulties. If the work is done outside, inclement weather can be a problem. More importantly, job analysts have difficulty observing some job tasks because they are rarely performed (e.g., police officer firing a weapon at a perpetrator) or are in settings that pose too much risk to the analyst (e.g., combat zones).

The last data collection method, lab or field simulations, are popular among analysts who favor physiological or biomechanical assessments. Simulations of specific physical job behaviors (e.g., dragging a dummy to simulate dragging a human out of harm's way) offer an advantage over observations because the analyst can observe the (simulated) behavior in a controlled or semi-controlled setting. Simulations also offer analysts the ability to evaluate tasks that are rarely performed on the job but are important. Depending on how the simulations are conducted, analysts may be able to estimate the physiological demands of the task in addition to how the task is performed. The main weakness of lab and field simulations is that they are resource-intensive, requiring specialized equipment and expertise. Also, simulations require the job incumbents to be at specific locations (e.g., a lab), requiring more time and resources.

Ideally, a job analyst will have the resources, expertise, and cooperation of the career field being analyzed to use any combination of job analytic methods in the toolkit. However, reality often means the analyst has to make decisions about which methods should and *can* be used in the given context.

Appendix B. Detailed Results from Movement Pattern Analyses

This appendix provides the detailed results of CPTs and average movement pattern ratings discussed in Chapter Four. Each movement pattern was rated on a 3-point scale (0 = not required, 1 = moderate amount, 2 = extensive amount) by experts within each specialty (CCT n=10; PJ n=5; SOWT n=4; TACP n=5).

Table B.1. Average Ratings on the Extent of Movement Pattern Used for CCT CPTs

Task	Balancing	Bending / Rotating	Carrying / Lifting	Climbing	Crouching / Squatting	Gripping / Handling	Pulling	Pushing	Running	Sprinting	Standing / Walking	Jumping
Carry casualties from immediate battle area to cover (e.g., use fireman's carry)	0.80	1.60	2.00	0.00	1.00	1.50	0.50	0.50	1.70	1.20	1.10	0.00
Participate in surface operations in urban environments with ruck load	1.00	0.90	1.80	0.90	0.90	0.80	0.00	0.00	0.90	0.80	1.20	0.40
Perform operational fast rope drop into a landing zone, e.g., from a helicopter (with no equipment, with LBE, weapon, and helmet, and with full combat gear)	1.00	0.50	0.00	0.00	0.40	2.00	0.00	0.00	0.90	0.00	1.20	0.00
Participate in surface operations in deserts with ruck load	0.60	0.90	1.80	0.50	1.00	0.90	0.00	0.00	0.90	1.20	1.20	0.00
Perform night infiltration with follow-on mission	0.80	1.40	1.70	0.90	1.30	1.30	0.40	0.40	1.80	1.40	1.40	0.80
Conduct day and night TAC missions	0.20	1.30	1.30	0.20	1.00	0.90	0.10	0.00	1.80	1.10	1.30	0.10
Perform immediate action drills (IADs)	0.60	1.40	0.60	0.10	1.40	1.80	0.20	0.20	1.30	1.30	1.40	0.10
Perform hand-to-hand combat	1.20	1.20	1.10	0.00	1.50	1.70	1.80	1.80	0.00	0.10	0.90	1.00
Drag casualties from immediate battle area to cover	0.80	1.80	1.60	0.00	1.80	1.90	1.70	0.00	1.20	1.20	1.40	0.00
Participate in surface operations in extreme cold environments with ruck load	0.20	1.30	1.80	0.10	1.30	1.30	0.40	0.40	1.20	0.40	1.60	0.00
Perform dismounted offensive operations	0.60	1.40	1.00	1.30	1.40	1.80	0.90	0.90	1.80	1.80	1.40	0.10
Perform overland infiltration or exfiltration operations	0.60	0.90	1.30	0.90	1.00	1.40	0.10	0.10	1.80	1.40	1.40	0.00
Perform night compass swim-scuba operations	0.00	0.70	0.60	0.00	0.00	1.50	0.00	0.10	0.00	0.00	0.00	0.00
Participate in surface operations at high altitudes above 8000 with ruck load	0.70	1.20	1.90	1.90	1.20	1.40	0.20	0.00	0.90	0.20	2.00	0.00

Task	Balancing	Bending / Rotating	Carrying / Lifting	Climbing	Crouching / Squatting	Gripping / Handling	Pulling	Pushing	Running	Sprinting	Standing / Walking	Jumping
Participate in surface operations in jungles with ruck load	1.20	1.70	1.90	1.20	1.30	1.80	0.00	0.00	0.00	0.00	1.60	0.00
Maneuver around, over, and under obstacles while in combat load	1.90	1.90	1.80	1.70	1.90	1.80	0.90	0.90	0.50	0.10	1.40	1.70

Table B.2. Average Ratings on the Extent of Movement Pattern Used for PJ CPTs

Task	Balancing	Bending / Rotating	Carrying / Lifting	Climbing	Crouching / Squatting	Gripping / Handling	Pulling	Pushing	Running	Sprinting	Standing / Walking	Jumping
Perform immediate action drills and execute tactics to break enemy contact (i.e., I'm up, he sees me, I'm down, REPEAT)	1.00	1.20	0.60	0.20	1.40	0.40	0.40	1.20	2.00	1.80	1.60	1.00
Perform medical treatment of wounds, injuries, and illnesses including combat casualty care, triage of mass casualties, life saving techniques, crisis action team functions, and JMAU duties	0.40	1.60	1.60	0.00	1.80	1.40	0.40	0.20	0.20	0.00	1.00	0.00
Carry or lift survivors from aircraft or vehicle wreckage	1.60	1.40	1.80	0.00	1.80	1.40	1.00	0.60	0.40	0.00	1.20	0.00
Drag casualties from immediate battle area to cover	0.80	1.60	1.40	0.20	1.20	1.00	1.80	0.40	0.60	0.80	0.60	0.40
Employ and operate extrication devices and equipment (e.g., axe, sledgehammer, crowbar)	0.40	0.60	0.40	0.00	1.00	1.00	1.20	1.00	0.00	0.00	0.60	0.00
Carry patients from immediate battle area (e.g., fireman carry)	1.60	1.80	1.80	0.00	1.60	1.40	0.60	0.40	0.80	0.80	1.00	0.00
Lift, load and unload equipment or personnel in litters onto trailers, vehicles, or aircraft	0.60	0.60	1.40	0.20	0.60	1.20	1.40	0.80	0.40	0.00	1.00	0.00
Perform technical rescue and recovery for confined space or imminent collapse rescue operations	0.00	1.00	0.60	0.40	1.20	1.00	1.40	0.40	0.40	0.00	0.40	0.00

Task	Balancing	Bending / Rotating	Carrying / Lifting	Climbing	Crouching / Squatting	Gripping / Handling	Pulling	Pushing	Running	Sprinting	Standing / Walking	Jumping
Perform downhill evacuation methods using low-angle evacuation, high-angle evacuation, buddy rappel, high-angle pick-off, improvised litters, suspension traverse method, rescue belay system, system know bypass	1.00	0.40	0.60	1.00	0.20	1.00	1.20	0.20	0.40	0.00	0.60	0.20
Maneuver body or litter over rocks or other obstructions during high angle rescue	0.80	1.00	1.40	1.40	0.40	1.20	1.40	0.60	0.00	0.00	0.80	0.00
Carry litter loaded with casualty for transfer or transload over flat or low-angle terrain	0.60	1.00	1.80	0.00	1.20	1.60	0.40	0.40	0.40	0.00	1.40	0.00
Perform water recoveries of personnel or materials	0.40	0.80	0.80	0.20	0.60	0.60	1.40	0.20	0.40	0.00	0.80	0.20
Climb rope ladder, caving ladder, or hook ladder to board aircraft or watercraft	0.80	0.40	0.00	1.80	0.00	1.80	1.80	0.00	0.00	0.00	0.00	0.00
Perform search and rescue (SAR) security team (SST) operations	0.60	1.00	0.60	0.40	1.00	0.60	0.40	0.20	0.80	1.00	1.40	0.20
Climb over walls and obstacles and crawl to stay in cover	1.00	0.60	0.00	1.60	1.00	1.00	1.40	1.00	0.40	0.60	0.80	1.00
Move over various terrain (e.g., jungle, mountain, desert, beach) with ruck load for extended surface operations, infiltration, or exfiltration	1.00	1.00	1.00	0.20	0.80	0.00	0.00	0.00	0.80	0.80	1.40	0.20
Prepare and execute open and closed-circuit dive operations including buddy rescue, underwater search and recovery, and dry-suit dive operations	0.40	0.60	0.40	0.00	0.20	0.60	0.60	0.00	0.00	0.00	0.00	0.00

Task	Balancing	Bending / Rotating	Carrying / Lifting	Climbing	Crouching / Squatting	Gripping / Handling	Pulling	Pushing	Running	Sprinting	Standing / Walking	Jumping
Perform movement or reconnaissance over difficult and varied terrain (e.g., glacier, mountain travel)	0.80	0.40	0.40	0.80	0.60	0.40	0.40	0.20	0.40	0.40	1.20	0.40
Swim with fins to unpack Ducks and RAMZ	0.20	1.00	0.00	0.00	0.00	0.00	0.60	0.00	0.00	1.00	0.00	0.00
Perform rope ascents and descents using ascending devices, friction knots, and roped party climbs	0.20	0.20	0.60	1.40	0.20	1.40	1.60	0.60	0.00	0.00	0.20	0.40
Perform underwater searches, such as circle line, clump line, parallel, or SONAR	0.40	0.20	0.20	0.00	0.00	0.60	0.60	0.40	0.00	0.00	0.00	0.00
Perform standard and emergency aircraft operations such as emergency crash and egress	0.80	1.40	0.60	0.60	0.40	0.60	1.00	0.80	1.60	0.80	0.80	0.00
Haul lines for recovery of personnel or equipment	0.40	0.40	1.20	0.20	0.80	1.80	1.80	0.00	0.00	0.00	0.60	0.00
Perform actions to survive, evade, resist, and escape the enemy including constructing shelters	0.40	0.40	0.60	0.40	1.00	0.60	0.60	0.40	0.60	0.40	1.20	0.60
Perform land search activities and procedures for personnel and equipment, such as contour, sweep land, or pace system searches	0.60	0.20	0.20	0.80	0.60	0.80	0.40	0.40	0.40	0.20	1.60	0.60
Perform assisted or unassisted climbs or descents, using different techniques (e.g., balance, chimney, ice wall, mantelshelf maneuver)	1.40	0.60	0.20	1.80	1.00	1.80	1.80	0.40	0.00	0.00	0.00	0.60

Table B.3. Average Ratings on the Extent of Movement Pattern Used for SOWT CPTs

Task	Balancing	Bending / Rotating	Carrying / Lifting	Climbing	Crouching / Squatting	Gripping / Handling	Pulling	Pushing	Running	Sprinting	Standing / Walking	Jumping
Move through, over, or around obstacles (except minefields)	1.50	1.25	0.75	1.50	1.25	0.75	0.75	0.75	1.25	1.50	1.50	1.50
Perform battle drills (e.g., react to indirect fire, evacuate a casualty, react to vehicle roll-over, enter and clear a room)	1.50	1.50	2.00	1.00	2.00	1.75	1.25	1.25	1.25	1.75	1.75	1.00
Create formations and use defensive fire techniques for overland infiltration and exfiltration operations	0.50	0.75	0.75	0.75	1.50	1.25	0.25	0.25	1.00	1.00	1.75	0.50
Move as a member of a fire team	1.00	0.75	1.50	0.50	1.00	1.00	0.50	0.50	1.00	1.25	1.75	0.75
Perform as a member of a patrol	0.50	0.75	1.25	0.25	0.75	0.75	0.25	0.25	0.75	0.50	1.50	0.50
Perform live fire immediate action drills (day or night)	1.25	1.00	1.25	1.00	1.50	1.25	0.75	0.75	1.00	1.50	1.50	1.00
Perform immediate action drills (IADs)	1.25	1.00	1.25	1.00	1.50	1.00	0.75	0.75	1.00	1.50	1.25	1.00
Perform land navigation (day or night)	1.00	0.50	1.25	0.50	0.50	0.50	0.25	0.25	0.25	0.25	1.75	0.50
Perform reconnaissance or surveillance activities	0.75	1.00	1.25	1.00	1.50	0.50	0.25	0.25	0.75	0.75	1.50	0.25
Perform movements under fire (direct or indirect) while mounted	0.75	0.75	0.75	0.25	1.00	1.25	0.75	0.50	0.25	0.25	0.75	0.25
Perform night movements	1.50	1.00	1.00	0.50	1.25	0.50	0.25	0.25	0.50	0.75	1.75	0.50
Perform movements under fire (direct or indirect) while dismounted	1.50	1.50	1.50	0.75	1.75	1.25	0.75	0.75	1.50	1.50	1.75	1.25
Perform small unit tactics (SMUTs) and patrolling	1.25	1.50	1.50	0.75	1.50	1.25	0.75	0.75	0.75	1.25	2.00	1.00

Table B.4. Average Ratings on the Extent of Movement Pattern Used for TACP CPTs

Task	Balancing	Bending / Rotating	Carrying / Lifting	Climbing	Crouching / Squatting	Gripping / Handling	Pulling	Pushing	Running	Sprinting	Standing / Walking	Jumping
React to enemy contact	0.40	1.40	0.40	0.80	1.80	0.60	0.40	0.40	1.60	1.80	0.20	1.00
React to improvised explosive devices	0.20	0.80	0.80	1.20	1.00	0.80	1.00	0.80	1.20	1.40	0.20	0.60
React to far/near ambush	0.60	1.20	0.80	0.80	1.80	0.60	0.40	0.40	1.60	1.80	0.40	1.00
React to direct/indirect fire	0.40	1.40	0.40	0.80	2.00	0.60	0.40	0.40	1.60	1.80	0.40	0.80
Evacuate injured personnel from vehicle	1.60	1.60	2.00	1.00	1.00	2.00	1.60	1.20	0.40	0.60	0.60	0.20
Navigate by foot during day or night operations using maps or compasses	1.40	0.20	0.60	0.40	0.60	0.00	0.00	0.00	0.20	0.00	1.80	0.20
Conduct dismounted operations	1.20	0.60	1.20	0.80	1.40	1.00	0.40	0.40	1.00	0.80	1.80	0.60
Perform foot marches	1.20	0.00	1.40	0.40	0.40	0.40	0.00	0.00	0.60	0.40	2.00	0.20
Perform small unit tactics	1.20	1.20	1.00	0.60	1.60	0.80	0.40	0.40	1.00	1.40	1.80	0.60
Break contact with the enemy	0.80	0.80	1.00	0.80	1.40	1.00	0.40	0.40	1.60	1.80	0.80	0.80
Perform offensive actions	0.80	1.00	1.00	0.80	1.20	1.00	0.40	0.40	1.60	1.40	1.00	0.80
Conduct infiltration, surface movement, and exfiltration functions with combat maneuver force	1.00	0.80	1.20	0.80	1.40	1.00	0.60	0.60	1.40	1.20	1.60	0.80
Perform team/squad movement techniques	1.00	0.80	1.20	0.80	1.60	1.00	0.40	0.40	1.20	1.20	1.60	0.60
Perform vehicle egress under roll-over conditions	1.20	1.80	1.20	1.60	1.40	1.80	1.40	1.40	0.00	0.20	0.20	0.60
Perform infiltrations or exfiltrations	1.00	0.80	1.20	1.00	1.40	1.00	0.40	0.40	1.20	1.00	1.40	1.00
Perform defensive actions	1.00	1.00	1.00	0.80	1.60	1.00	0.40	0.60	1.60	1.40	1.20	0.80
Enter/clear a building	1.60	1.20	0.60	1.20	1.20	1.00	0.60	0.60	0.60	0.40	1.80	0.60

Task	Balancing	Bending / Rotating	Carrying / Lifting	Climbing	Crouching / Squatting	Gripping / Handling	Pulling	Pushing	Running	Sprinting	Standing / Walking	Jumping
Perform survival, evasion, resistance, and escape (SERE) activities	1.60	1.40	1.00	1.40	1.80	1.20	1.40	1.20	1.60	1.60	1.60	1.40
Perform individual movement tactics	1.60	1.40	1.40	1.00	1.80	1.00	0.40	0.40	1.00	1.20	1.80	0.80
Perform assault zone operations	1.00	1.20	1.00	0.80	1.60	1.00	0.80	0.60	1.20	1.20	1.40	1.00

Appendix C. Average Task Ratings from Focus Groups

This appendix provides the average (mean) ratings of task importance, intensity, frequency, and duration based on the focus group surveys. The rating scale for these dimensions are provided below. Tables C.1–C.4 include only means for which at least 60 percent of the sample (i.e., total number of focus group participants in the specialty) provided affirmative responses (i.e., no missing or "not applicable" responses). SOWT, for example, had 16 total participants, meaning that a rating needed nine affirmative responses for a mean to be calculated. If a particular rating did not meet the threshold sample size, we left the cell blank in the table. We applied the same criterion to the other three AFSs. The sample size varied across AFS; CCT (n=57), PJ (n=28), SOWT (n=16), TACP (n=114).

Focus Group Rating Dimensions

Importance

How important are these tasks to effective mission performance? Although you may not perform each task in every mission, please consider the importance of the task when it is performed.

1 = Not Important
2 = Somewhat Important
3 = Moderately Important
4 = Very Important
5 = Crucial

Frequency

How often have you performed this task in your missions over the past 2 years? Please note, we are concerned only with your performance of this task during an *operational mission*, not during a training mission.

1 = Never Performed (e.g., received training but never had to perform in a mission)
2 = Seldom Performed (e.g., less than 25% of missions)
3 = Occasionally Performed (e.g., approximately 50% of missions)
4 = Often Performed (e.g., more than 75% of missions)
5 = Always Performed (e.g., on every mission, multiple times during a mission)

Duration

In general, about how long does it take for you to perform this task before you can either rest or move on to a new task?

1 = 0 to 2 minutes
2 = 3 to 30 minutes
3 = 31 minutes to 1 hour
4 = 1 to 2 hours
5 = More than 2 hours

Intensity

How much physical effort is required to successfully complete this task? Physical effort is defined by strength, endurance, or movement quality (i.e., balance, flexibility).

1 = Very, Very Light
2 = Very Light
3 = Light
4 = Somewhat Hard
5 = Hard
6 = Very Hard
7 = Very, Very Hard

Table C.1. Average Ratings for CCT Tasks

Task	Importance	Intensity	Frequency	Duration
Participate in river or stream crossings	3.09	3.56	2.15	1.95
Rig or derig equipment for air drops, e.g. water craft (CRRCs/RAMZs)	3.43	3.18	2.24	2.47
Participate in casting or recovery operations, other than cast master activities	3.35	3.31	1.78	2.59
Locate CAS targets	4.88	3.37	3.92	3.50
Don or doff chemical warfare personal protective equipment (PPE)	2.63	2.13	1.57	1.65
Perform operational helicopter air-land employments	4.52	3.22	3.04	2.36
Perform operational hoist insertions or extractions	3.33	3.05	1.67	2.04
Carry casualties from immediate battle area to cover (e.g., use fireman's carry)	4.51	5.52	2.35	2.27
Perform base camp movement	3.63	4.28	2.27	4.22
Participate in surface operations in urban environments with ruck load	4.45	5.11	3.58	4.60

Task	Importance	Intensity	Frequency	Duration
Perform operational fast rope drop into a landing zone, e.g., from a helicopter (with no equipment, with LBE, weapon, and helmet, and with full combat gear)	4.07	4.11	2.12	1.78
Operate all-terrain vehicles (ATVs)	4.12	2.82	3.42	3.91
Perform fast-roping, roping, caving ladder, or rappelling operations	3.85	4.04	1.96	1.89
Perform self-aid or buddy care procedures	4.78	3.44	2.30	2.55
Participate in surface operations in deserts with ruck load	4.44	5.25	3.14	4.45
Enforce perimeter security during patrols and bivouacs	4.46	3.34	3.50	4.09
Perform helicopter short tactical airborne operation (STABO), special patrol insertion/extraction system (SPIES), or fast-rope insertion/extraction system (FRIES) operations	3.61	3.37	1.73	2.05
Establish ground marked release points	3.83	2.17	2.12	2.13
Perform night Infiltration with follow-on mission	4.70	4.71	2.87	4.55
Conduct day and night TAC missions	4.75	4.19	3.71	4.54
Perform immediate action drills (IADs)	4.63	4.09	2.96	2.89
Perform operational fixed-wing air-land employments	4.42	2.82	2.49	3.06
Signal aircraft	4.30	1.89	3.33	1.96
Perform assault zone surveying activities	4.38	3.07	2.75	3.61
Perform hand-to-hand combat	4.11	4.78	1.65	2.16
Perform rapid extraction deployment system (REDS) riggings	3.08	2.85	1.39	2.32
Perform explosive ordnance recognition (EOR) procedures	4.19	2.60	2.30	2.73
Perform navigation using global positioning system (GPS)	4.77	2.27	4.48	4.05
Issue light gun signals	3.91	1.29	1.53	1.55
Perform operational dive employments with fins	3.85	4.22	1.78	3.51
Perform day compass swim-scuba operations	3.73	4.21	1.76	3.42
Perform day or night combat convoy operations	4.48	3.47	3.54	4.45
Participate in mountaineering operations using technical rope systems	3.48	4.02	1.51	3.38
Assess assault zones	4.57	3.15	2.76	3.36
Perform operational small boat employments	3.70	3.14	1.76	3.40
Perform mounted offensive operations	4.48	3.77	3.08	4.21
Perform day infiltration	4.39	3.88	3.50	3.95
Drag casualties from immediate battle area to cover	4.78	5.28	2.12	2.02
Identify and report controlled firing areas	4.28	2.50	2.20	2.31
Participate in surface operations in extreme cold environments with ruck load	4.11	5.13	2.15	4.28
Gather or report intelligence data using photographic equipment	4.07	2.37	2.27	3.25
Perform operational parachute employments	4.47	3.35	1.88	2.63
Fire weapons at night	4.84	3.69	3.06	3.47

Task	Importance	Intensity	Frequency	Duration
Perform parachute Jumps	4.58	3.54	2.14	2.88
Perform dismounted offensive operations	4.64	4.71	3.09	4.14
Perform or set up site security	4.50	3.45	3.25	3.76
Handle prisoners of war (POWs)	3.92	3.16	2.24	3.34
Perform camouflage procedures	3.70	1.86	2.33	2.49
Perform overland infiltration or exfiltration operations	4.57	4.73	3.17	4.36
Perform night compass swim-scuba operations	4.09	4.35	1.54	3.65
Participate in surface operations at high altitudes above 8000 with ruck load	4.57	5.98	2.52	4.58
Construct field fortifications or shelters	3.89	3.49	2.06	3.63
Perform scuba equipment ditching or donning procedures	3.91	3.54	1.55	1.66
Perform protective mask firings	3.35	2.80	1.45	1.93
Set up or tear down tents or shelters	3.02	2.77	1.96	2.77
Participate in watercraft operations, such as over-the-horizon or riverine operations	3.81	3.30	1.48	3.57
Execute maneuvers using hand or arm signals	4.32	2.14	2.87	2.58
Perform cardiopulmonary resuscitation (CPR) procedures	4.41	3.02	1.56	1.86
Establish equipment security at mission locations	4.46	2.64	3.08	3.10
Participate in surface operations in jungles with ruck load	4.27	5.00	1.43	4.25
Gather or report intelligence data, other than using photographic equipment	4.44	2.45	2.73	3.38
Report BDAs during CAS operations	4.72	2.35	3.37	2.26
Perform operational helo cast employments	3.90	3.38	1.65	2.33
Perform reconnaissance or surveillance activities	4.56	3.77	2.69	4.40
Operate tactical vehicles	4.61	2.79	3.62	4.31
Perform terminal control	4.82	3.41	3.40	3.91
Maneuver around, over, and under obstacles while in combat load	4.68	5.13	3.74	3.87

Table C.2. Average Ratings for PJ Tasks

Task	Importance	Intensity	Frequency	Duration
Perform standard and emergency aircraft operations such as emergency crash and egress	4.21	4.15	1.46	1.85
Perform assisted or unassisted climbs or descents, using different techniques (e.g., balance, chimney, ice wall, mantelshelf maneuver)	4.00	4.93	1.61	2.96
Swim with fins to unpack Ducks and RAMZ	4.39	4.54	1.64	2.45
Carry or lift survivors from aircraft or vehicle wreckage	4.89	5.57	2.46	2.20
Don, adjust, and doff equipment (e.g., SCUBA, parachutes, personal protective equipment)	4.32	2.93	3.32	2.04
Carry litter loaded with casualty for transfer or transload over flat or low-angle terrain	4.54	5.04	2.93	2.48
Perform airfield seizure for strobe man duties, RATT operations, joint casualty collection point (JCCP) operations	3.68	3.80	1.36	3.05
Climb over walls and obstacles and crawl to stay in cover	4.46	4.82	1.93	2.29
Perform hand-to-hand combat	3.12	5.43	1.04	1.87
Drag casualties from immediate battle area to cover	4.82	5.89	1.89	1.87
Construct tools to aid in personnel or equipment survival or recovery (e.g., improvised litters, mechanical advantages)	4.29	3.32	1.64	2.56
Haul, pull, or lift equipment and bundles for deployment	3.82	4.71	2.57	2.16
Move over various terrain (e.g., jungle, mountain, desert, beach) with ruck load for extended surface operations, infiltration, or exfiltration	4.46	5.29	2.07	3.67
Swim with fins over extended periods to execute open and closed-circuit operations, infiltration, or exfiltration	3.70	4.81	1.37	3.27
Prepare and execute open and closed-circuit dive operations including buddy rescue, underwater search and recovery, and dry-suit dive operations	4.46	4.36	1.43	3.33
Perform immediate action drills and execute tactics to break enemy contact (i.e., I'm up, he sees me, I'm down, REPEAT)	4.93	5.48	1.46	2.12
Perform security procedures to control and maintain safety of prisoners, crowds, and survivors	4.18	3.22	2.04	2.81
Perform tactics, techniques, and procedures to enforce and secure sites, aircrafts, and sensitive equipment	4.64	3.59	1.93	3.37
Perform water recoveries of personnel or materials	4.50	4.56	1.75	3.59
Climb rope ladder, caving ladder, or hook ladder to board aircraft or watercraft	4.50	5.00	1.79	1.57
Perform tandem parachute operations for equipment or personnel	3.13	4.55	1.18	1.82
Perform infiltration using rappel or fast rope techniques	4.64	3.74	1.82	1.26
Lift, load and unload equipment or personnel in litters onto trailers, vehicles, or aircraft	4.64	4.96	3.04	1.77

Task	Importance	Intensity	Frequency	Duration
Perform medical treatment of wounds, injuries, and illnesses including combat casualty care, triage of mass casualties, life saving techniques, crisis action team functions, and JMAU duties	4.93	4.61	3.25	2.82
Employ and operate extrication devices and equipment (e.g., axe, sledgehammer, crowbar)	4.79	5.67	1.79	2.74
Perform rope ascents and descents using ascending devices, friction knots, and roped party climbs	4.39	4.65	1.43	2.56
Use parachute riser manipulation in emergency procedures	4.71	3.96	1.32	1.33
Maneuver body or litter over rocks or other obstructions during high angle rescue	4.57	5.48	1.61	2.93
Perform DZ controller duties	3.41	2.04	1.54	2.56
Haul lines for recovery of personnel or equipment	4.21	5.04	1.46	2.67
Perform technical rescue and recovery for confined space or imminent collapse rescue operations	4.64	5.33	1.29	3.52
Carry patients from immediate battle area (e.g., fireman carry)	4.71	5.74	1.57	1.81
Recover casualties using standard and improvised equipment including tag lines, forest penetrators, horse collars, navy slings, and hauling lines	4.29	3.85	1.64	1.89
Perform mounted vehicle movement (e.g., tactical wheeled vehicles, tactical watercraft)	3.68	3.04	1.50	3.22
Rig or derig deployment equipment, including RAMZ, all-terrain vehicles (ATVs), or motorcycles	4.14	3.15	1.50	2.30
Perform parachute jumps (e.g., static line, free fall, high attitude)	4.54	3.52	1.57	1.89
Operate, transport, and safeguard weapons, such as grenade launchers, handguns, crew-served, or rifles	4.36	2.86	3.36	3.57
Perform search and rescue (SAR) security team (SST) operations	4.50	4.00	1.85	3.77
Perform water-based search operations using boats or diver tow procedures	3.79	3.62	1.36	3.37
Perform actions to survive, evade, resist, and escape the enemy including constructing shelters	4.21	4.63	1.14	4.33
Perform movement or reconnaissance over difficult and varied terrain (e.g., glacier, mountain travel)	4.43	5.33	1.71	4.22
Perform underwater searches, such as circle line, clump line, parallel, or SONAR	4.25	4.19	1.39	3.30
Perform downhill evacuation methods using low-angle evacuation, high-angle evacuation, buddy rappel, high-angle pick-off, improvised litters, suspension traverse method, rescue belay system, system know bypass	4.64	4.63	1.57	3.26
Perform water, river, or mountain stream crossings	3.93	4.07	1.21	2.26
Launch, raise, and lower watercrafts to and from other vessels	3.57	3.59	1.21	2.30
Operate watercrafts including alternate rescue craft (ARC), inflatable watercraft, rigid hull watercraft, and military amphibious reconnaissance system (MARS) engine	4.11	3.15	1.57	3.04

Task	Importance	Intensity	Frequency	Duration
Perform land search activities and procedures for personnel and equipment, such as contour, sweep land, or pace system searches	4.21	3.89	1.63	3.70

Table C.3. Average Ratings for SOWT Tasks

Task	Importance	Intensity	Frequency	Duration
Establish and control helicopter landing zones (HLZs) or pickup zones	3.44	3.07	2.38	2.73
Control dismounted movement formations	3.40	4.42	2.20	4.00
Move through, over, or around obstacles (except minefields)	4.00	4.69	3.38	3.44
Perform day or night rotary-wing infiltration or exfiltration activities	3.88	3.56	2.38	3.09
Participate in watercraft operations, such as over-the-horizon or riverine operations	2.94	3.18	1.73	3.46
Demonstrate the principles of survival to operate in night environments using night vision devices	4.44	3.38	3.31	4.56
Demonstrate the principles of survival to operate in nuclear, biological and chemical environments	2.93		1.31	
Perform dive supervisor activities for open or closed circuit dives				
Perform bivouac and patrol base duties	3.87	3.82	2.47	4.71
Perform open circuit dives (day or night)				
Perform concealed movement	3.81	4.67	2.50	4.53
Perform caving ladder activities	2.47	4.90	1.47	
Perform drop, landing, pick-up, or extraction zone activities, including establishing zones	3.88	3.44	2.31	3.53
Perform danger area procedures	3.75	2.92	2.38	3.08
Perform environmental reconnaissance	4.56	3.19	4.06	3.63
Perform battle drills (e.g., react to indirect fire, evacuate a casualty, react to vehicle roll-over, enter and clear a room)	4.63	5.17	2.50	3.20
Create formations and use defensive fire techniques for overland infiltration and exfiltration operations	4.13	4.40	2.44	3.70
Enforce perimeter security during patrols and bivouacs	4.25	3.27	2.88	4.57
Move as a member of a fire team	4.25	4.23	2.94	4.13
Operate dissimilar parachute systems	3.19		1.31	
Perform as a member of a patrol	4.44	4.20	3.31	4.33
Perform live fire immediate action drills (day or night)	4.56	4.29	2.31	2.64
Execute maneuvers using hand or arm signals	3.44	2.69	1.88	2.65
Perform day or night fixed-wing infiltration or exfiltration activities	3.63	2.86	2.13	2.79
Assess and respond to threats (escalation of force): Employ progressive levels of individual force when confronting civilians, challenge persons entering your area	3.94	3.71	2.13	2.36
Perform immediate action drills (IADs)	4.38	4.03	2.31	2.50
Construct hasty or defensive fighting positions	3.94	4.47	2.19	2.53
Perform helocast: day helocast, night helocast, cast-master duties	2.94		1.27	
Perform water jump (day or night)	2.75	3.50	1.27	2.70

Task	Importance	Intensity	Frequency	Duration
Operate unmanned aerial systems (UASs)	4.07	2.36	2.71	3.29
Employ combat rubber raiding crafts (CRRCs) or rigging alternate method Zodiacs (RAMZs)	2.63	0.00	1.21	0.00
Perform equipment parachute jumps during day or night operations	3.56	0.00	1.33	0.00
Perform explosive ordnance reconnaissance	3.33	3.32	1.93	3.50
Perform land navigation (day or night)	4.13	4.47	2.56	4.43
Perform free fall parachute jumps	3.31		1.00	
Perform closed circuit dives (day or night)	1.90			
Perform fast rope infiltration or exfiltration (FRIES) operations	3.63	4.45	1.47	2.18
Perform reconnaissance or surveillance activities	4.38	4.25	2.81	4.75
Perform small boat operations: open water CRRC, kayak/canoe, rigid hull	2.80		1.15	
Perform survival, evasion, resistance, and escape (SERE) activities	3.88		1.38	
Perform operational fixed-wing airland employments	3.50	2.58	1.87	2.92
Prepare equipment for amphibious operations	2.73		1.21	
Perform scout swimming	2.54		1.30	
Perform surface and subsurface swimming	2.47		1.14	
Rig or derig ATVs, motorcycles, or waverunners for airdrops	3.27		1.36	
Respond to depleted uranium	2.00			
Perform rope ladder activities	3.27		1.21	
Perform operational hoist insertions or extractions (on land or water, day or night)	3.00		1.14	
Use camouflage, cover and concealment; field hygiene techniques; night observation illumination; protective mask	3.87	2.54	2.60	3.77
Perform sniper countermeasures	3.08			
React to unexploded ordnance hazards	3.57	2.60	1.85	2.70
Perform rappelling activities (tower, helo)	3.29		1.31	
Perform operations using night vision devices	4.40	3.73	3.00	4.60
Perform mountaineering activities	3.80	5.42	2.00	4.67
Perform movements under fire (direct or indirect) while mounted	4.56	4.60	2.31	3.32
Perform night movements	4.19	4.28	2.56	4.60
Perform perimeter defense activities	4.13	3.47	2.63	4.07
Perform search dives (day or night)	2.00			
Perform movements under fire (direct or indirect) while dismounted	4.38	5.11	2.25	3.14
React to flares	2.46		1.42	
Perform small boat navigation during over the horizon operations, riverine operations, or inter-coastal operations	2.71		1.15	

Task	Importance	Intensity	Frequency	Duration
Prepare equipment for subsurface operations	2.00			
Perform weather reconnaissance	4.75	3.19	4.25	3.69
Search vehicles in a tactical environment	3.14	2.70	2.08	2.60
Perform small unit tactics (SMUTs) and patrolling	4.27	4.27	2.88	4.40
Rig or derig combat rubber raiding crafts (CRRCs) or rigging alternate method Zodiacs (RAMZs) for airdrops	2.67		1.07	
Select temporary fighting positions	4.20	3.63	2.40	3.03

Table C.4. Average Ratings for TACP Tasks

Task	Importance	Intensity	Frequency	Duration
Break contact with the enemy	4.46	5.37	2.29	2.37
Conduct dismounted operations	4.57	4.86	3.59	4.59
Conduct Rally Point procedures	3.55	3.05	2.34	
Cross danger areas	4.26	3.96	2.96	2.40
Dismount a vehicle	3.79	2.68	3.50	1.49
Perform Air Force Combative	3.10	4.78	1.32	
Perform foot marches	4.52	5.40	3.55	4.56
Perform individual movement tactics	4.16	4.26	3.05	3.19
Perform offensive actions	4.44	4.59	2.72	3.35
Perform operations using night vision devices	4.58	3.92	3.25	4.33
Perform small unit tactics	4.47	4.30	2.96	3.85
Perform team/squad movement techniques	4.38	4.15	3.13	3.82
Perform mounted and dismounted navigation	4.52	3.98	3.21	4.11
React to direct/indirect fire	4.80	5.01	2.57	2.34
React to enemy contact	4.85	5.43	2.46	2.61
React to far/near ambush	4.83	5.55	2.11	2.57
React to flares	3.52	3.15	1.27	
React to improvised explosive devices	4.83	4.69	2.22	2.59
Evacuate injured personnel from vehicle	4.69	5.20	1.70	
Perform vehicle egress under roll-over conditions	4.37	4.68	1.34	
Perform vehicle recovery operations	3.72	3.88	1.88	
Conduct infiltration, surface movement, and exfiltration functions with combat maneuver force	4.44	4.49	3.07	3.67
Depart and reenter friendly lines	3.88	2.99	3.19	2.35
Perform infiltrations or exfiltrations	4.34	4.14	3.11	2.92
Participate in air mobile procedures	3.95	3.56	2.66	
Perform assault zone operations	4.08	4.00	2.36	
Perform immediate-action drills	4.53	3.76	2.30	
Perform combat water survival procedures	3.10		1.07	
Perform survival, evasion, resistance, and escape (SERE) activities	4.20		1.08	
Establish and control helicopter landing zones (HLZs) or pickup zones	4.13	3.43	2.24	
Mark drop zones	3.78	3.01	1.95	
Mark landing zones	3.70	2.96	1.99	
Perform drop zone operations	3.55	3.06	1.81	
Perform drop zone control team procedures	3.45		1.66	
Perform helicopter landing zone operations	3.79	3.19	2.22	
Coordinate fast rope infiltration or exfiltration system (FRIES) operations with Army or other service personnel	3.16			

Task	Importance	Intensity	Frequency	Duration
Perform alternate insertions or extractions, other than equipment or tactical helicopter rappellings	3.25		1.82	
Perform FRIES operations	3.17			
Perform helicopter insertions or extractions	4.15	3.92	2.88	2.28
Perform tactical helicopter rappellings	3.15			
Construct an improvised litter	3.85	3.60	1.48	
Enter/clear a building	4.25	4.72	2.22	
Participate in Army field exercises or rehearsals	4.04	3.50	3.44	4.13
Position equipment for operational use	4.52	3.01	4.22	3.22
Use field gear/equipment	4.77	3.52	4.52	3.99
Navigate by foot during day or night operations using maps or compasses	4.63	4.35	2.92	4.16
Navigate by foot using global positioning system (GPS) equipment	4.67	3.99	3.40	4.20
Perform predeployment reconnaissance surveys	3.36	2.32	2.43	
Construct defensive fighting positions	3.68	4.61	1.93	
Perform defensive actions	4.32	4.42	2.40	3.06
Perform or set up sites security	4.23	3.68	2.72	3.21
Perform patrol base procedures (PBPs)	3.97	3.58	2.39	
Perform perimeter security	3.98	3.43	2.45	
Prepare sites at deployed locations, such as digging defensive firing positions	3.62	4.49	1.87	
Participate in search and rescue missions	3.58		1.51	
Perform explosive ordnance reconnaissance	3.37	3.37	1.99	
Perform patrol element team responsibilities	3.83	3.53	2.29	
Participate in waterborne operations, such as helo-castings	2.60			
Perform equipment parachute jumps	3.51			
Perform jumpmaster aircraft inspections	3.21			
Perform jumpmaster operations, other than jumpmaster aircraft inspections or jumpmaster military free-fall operations	3.24			
Perform military free-fall operations, other than jumpmaster military free-fall operations	3.10			
Perform static-line parachute jumps	3.48			
Perform tactical parachute jumps during day operations	3.40			
Perform tactical parachute jumps during night operations	3.51			
Perform water parachute jumps	3.09			
Pick up, deliver, or store equipment, tools, parts, or supplies	3.36	2.91	2.74	
Perform river crossings	3.60		1.74	
Perform drown proofing techniques	3.31		1.23	

Bibliography

Air Force Enlisted Classification Directory (AFECD), *The Official Guide to the Air Force Enlisted Classification Codes*, Washington, D.C.: Headquarters, U.S. Air Force, August 1, 2012.

Air Force Special Tactics (24 SOW), "Tactical Air Control Party," web page, no date. As of January 8, 2015: http://www.24sow.af.mil/Units/TacticalAirControlParty.aspx

Algera, J. A., and M. A. M. Greuter, "Job Analysis," in P. J. D. Drenth, H. Thierry, and C. J. de Wolff, eds., *A Handbook of Work and Organizational Psychology: Personnel Psychology*, 2nd ed., East Sussex, UK: Psychology Press LTD, 1998, pp. 141–164.

American College of Sports Medicine, *ACSM's Guidelines for Exercise Testing and Prescription* (8th ed.), Philadelphia: Lippincott Williams & Wilkins, 2010.

Annett, J., K. D. Duncan, R. B. Stammers, and M. J. Gray, *Task Analysis*, Training Information Paper No. 6, London: Her Majesty's Stationary Office (HMSO), 1971.

Arvey, R.D. and K. R. Murphy, "Performance Evaluation in Work Settings," *Annual Review of Psychology*, Vol. 49, 1998, pp. 141–168.

Arvey, Richard D., Steven M. Nutting, and Timothy E. Landon, "Validation Strategies for Physical Ability Testing in Police and Fire Settings," *Public Personnel Management*, Vol. 21. No. 3, 1992, pp. 301–313.

Åstrand, P-O, K. Rodahl, H. A. Dahl, and S. B. Strømme, *Textbook of Work Physiology: Physiological Bases of Exercise*, 4th ed., Champaign, Il.: Human Kinetics, 2003.

Ayoub, M. M., B. C. Jiang, J. L. Smith, J. L. Selan, and J. W. McDaniel, "Establishing a Physical Criterion for Assigning Personnel to U.S. Air Force Jobs," *American Industrial Hygiene Association Journal*, Vol. 48, No. 5, 1987, pp. 464–470.

Baker, T. A., and D. L. Gebhardt, "The Assessment of Physical Capabilities in the Workplace," in N. Schmitt, ed., *The Oxford Handbook of Personnel Assessment and Selection*, New York: Oxford University Press, 2012, pp. 274–296.

Bernard, B. P., ed., *Musculoskeletal Disorders and Workplace Factors: A Critical Review of Epidemiological Evidence for Work-Related Musculoskeletal Disorders of the Neck, Upper Extremity, and Low Back*, Cincinnati, Ohio: National Institute for Occupational Safety and Health, DHHS (NIOSH) Publication No. 97-141, July 1997.

Bell, D. R., K. M. Guskiewicz, M. A. Clark, and D. A., "Systematic Review of the Balance Error Scoring System," *Sports Health: A Multidisciplinary Approach*, Vol. 3, No. 3, 2011, pp. 287–295.

Bilzon, J. L. J., E. G. Scarpello, E. Bilzon, and A. J. Allsopp, "Generic Task-Related Occupational Requirements for Royal Naval Personnel," *Occupational Medicine*, Vol. 52, No. 8, 2002, pp. 503–510.

Bilzon, J. L., E. G. Scarpello, C. V. Smith, N. A. Ravenhill, and M. P. Rayson, "Characterization of the Metabolic Demands of Simulated Shipboard Royal Navy Fire-Fighting Tasks," , Vol. 44, No. 8, 2001, pp. 766–780.

Blakley, B. R., M. A. Quinones, M. S. Crawford, and I. A. Jago, "The Validity of Isometric Strength Tests," *Personnel Psychology*, Vol. 47, No. 2, 2004, pp. 247–274.

Borg, G., "Perceived Exertion as an Indicator of Somatic Stress," *Scandinavian Journal of Rehabilitation Medicine*, Vol. 2, No. 2, 1970, p. 92.

Borg, G., *Borg's Perceived Exertion and Pain Scales*, Champaign, Il.: Human Kinetics, 1998.

Bozic, P. R., N. R. Pazin, B. B. Berjan, N. M. Planic, and I. D. Cuk, I. D., "Evaluation of the Field Tests of Flexibility of the Lower Extremity: Reliability and the Concurrent and Factorial Validity," *Journal of Strength and Conditioning Research*, Vol. 24, No. 9, 2010, pp. 2523–2531.

Callinan, M., and I. T. Robertson, "Work Sample Testing," *International Journal of Selection and Assessment*, Vol. 8, No. 4, 2000, pp. 248–260.

Campbell, John P., "Modeling the Performance Prediction Problem," in W. C. Borman, D. R. Ilgen, and R. J. Klimoski, eds., *Handbook of Psychology: Industrial and Organizational Psychology*, Hoboken, N.J.: John Wiley & Sons, Inc., 1990, pp. 687–731.

Career Force Education and Training Plan (CFETP) 1W0XX, Change 5, *Weather*, Washington, D.C.: Headquarters, U.S. Air Force, March 15, 2012.

Christal, R. E., *The United States Air Force Occupational Research Project*, Lackland, Air Force Base, Tex.: U.S. Air Force Human Resources Laboratory, Occupational Research Division, AFHRL-TR-73-75, 1974.

Coburn, J. W., "Measuring Power," *Strength and Conditioning Journal*, Vol. 34, No. 6, 2012, pp. 25–28.

Davis, D. S., R. O. Quinn, C. T. Whiteman, J. D. Williams, and C. R. Young, "Concurrent Validity of Four Clinical Tests Used to Measure Hamstring Flexibility," *Journal of Strength and Conditioning Research*, Vol. 22, No. 2, 2008, pp. 583–588.

Department of the Air Force, "Tactical Air Control Party AFSC 1C4X1," Career Field Education and Training Plan 1C4X1 Parts I and II, November 15, 2009, Para. 4.2.

Desmond, R. E., and D. J. Weiss, "Supervisor Estimation of Abilities Requires in Jobs," *Journal of Vocational Behavior*, Vol. 3, No. 2, pp. 181–194.

Desmond, R. E., and D. J. Weiss, "Worker Estimation of Ability Requirements of Their Jobs," *Journal of Vocational Behavior*, Vol. 7, No. 1, pp. 13–28.

Dubois, Cathy L. Z., Paul R. Sackett, Sheldon Zedeck, and Larry Fogli, "Further Exploration of Typical and Maximum Performance Criteria: Definitional Issues, Prediction, and White-Black Differences," *Journal of Applied Psychology,* Vol. 78, No. 2, 1993, pp. 205–211.

Dziados, J. E., A. I. Damakosh, R. P. Mello, J. A. Vogel, L. Kenneth, and J. Farmer, *Physiological Determinants of Load Bearing Capacity*, Technical Report T19-87, Natick, Mass.: U.S. Army Research Institute of Environmental Medicine, 1987.

Eisinger, G. C., P. Wittels, R. Enne, M. Zeilinger, W. Rausch, Y. Hölzl, T., G. Dorbner, and N. Bachl, "Chapter 6: Evidence-Based Job Analysis and Methodology to Determine Physical requireaments of Special Miliary Occupations," in *Optimizing Operational Physical Fitness*, NATO Reseach and Technology Organisation, 2009.

Fine, S. A., and W. W. Wiley, *An Introduction to Functional Job Analysis*, Kalamazoo, Mich.: Upjohn Institute for Employment Research, 1971.

Flanagan, J. C., "The Critical Incident Technique," *Psychological Bulletin*, Vol. 51, No. 4, 1954, pp. 327–358.

Fleishman, E. A., "Dimensional Analysis of Movement Reactions," *Journal of Experimental Psychology*, Vol. 55, No. 5, 1958, pp. 438–453.

Fleishman, E. A., *Structure and Measurement of Physical Fitness*, Englewood Cliffs, N.J.: Prentice-Hall, 1964.

Fleishman, E. A., *Development of Ability Requirements Scales for the Analysis of Bell System Jobs*, Bethesda, Md.: Management Research Institute, 1975.

Fleishman, E. A., and J. C. Hogan, *Taxonomic Method for Assessing the Physical Requirements of Jobs: The Physical Abilities Approach*, Washington, D.C.: Advanced Research Resources Organization, ARRO Technical Report 3012/R78-6, 1978.

Fleishman, E. A., and M. D. Mumford, "Ability Requirement Scales," in S. Gael, ed., *The Job Analysis Handbook for Business, Industry, and Government: Volume II*, New York: John Wiley & Sons, 1988, pp. 917–935.

Fleishman, E. A., and M. K. Quaintance, *Taxonomies of Human Performance*, New York: Academic Press, 1984.

Gamble, R. P., A. B. Stevens, H. Mcbrien, A. Black, G. W. Cran, and C. A. G. Boreham, "Physical-Fitness and Occupational Demands of the Belfast Ambulance Service," *British Journal of Industrial Medicine*, Vol. 48, No. 9, 1991, pp. 592–596.

Gebhardt, Deborah L., and Todd A. Baker, "Physical Performance," in John C. Scott and Douglas H. Reynolds, eds., *Handbook of Workplace Assessment: Evidence-Based Practices for Selecting and Developing Organizational Talent*, San Francisco, Calif..: Jossey-Bass, 2010, pp. 165–196.

Gilliland, Stephen W., " The Perceived Fairness of Selection Systems: An Organizational Justice Perspective," *The Academy of Management Review*, Vol. 18, No. 4, 1993, pp. 694–734.

Göbel, M., "Electromyography (EMG)," in N. Stanton, A. Hedge, K. Brookhuis, E. Salas, and H. Hendrick, eds., *Handbook of Human Factors and Ergonomics Methods*, Boca Raton, Fl.: CRC Press, 2005, pp. 19-1–19-8.

Gutman, Arthur, "Legal Constraints on Personnel Selection Decisions," in Neal Schmitt, ed., *The Oxford Handbook of Personnel Assessment and Selection*, New York: Oxford University Press, 2012, pp. 686–720.

Harwood, Georgina E., Mark P. Rayson, and Alan M. Neville, "Fitness, Performance and Risk of Injury in British Army Officer Cadets," *Military Medicine*, Vol. 164, 1999, pp. 428–434.

Hashim, A., "Objectivity, Reliability, and Validity of the 90° Push-Ups Test Protocol Among Male and Female Students of Sports Science Program," *Journal of Physical Education and Sport*, Vol. 12, No, 1, 2012, pp. 103–106.

Henderson, Norman D., Michael W. Berry, and Tomislov Matic, "Field Measures of Strength and Fitness Predict Firefighter Performance on Physically Demanding Tasks," *Personnel Psychology*, Vol. 60, No. 2, 2007, pp. 431–473.

Hodgdon, James A., and Andrew S. Jackson, "Physical Test Validation for Job Selection," in Stefan Constable and Barbara Palmer, eds., *The Process of Physical Fitness Standards Development*, Wright-Patterson Air Force Base, Ohio: Human Systems Information Analysis Center, 2000, pp. 139–177.

Hogan, J., "Structure of Physical Performance in Occupational Tasks," *Journal of Applied Psychology*, Vol. 76, No. 4, 1991, pp. 495–507.

Hogan, J. C., "Physical Abilities," in M. D. Dunnette and L. M. Hough, eds., *Handbook of Industrial and Organizational Psychology*, Vol. 2, 2nd ed., Palo Alto, Calif.: Consulting Psychologists Press, 1991, pp. 753–831.

Hogan, J. C., and E. A. Fleishman, "An Index of Physical Effort Required in Human Task Performance," *Journal of Applied Psychology*, Vol. 64, No. 2, 1979, pp. 197–204.

Hogan, J. C., G. D. Ogden, D. L. Gebhardt, and E. A. Fleishman, "Reliability and Validity of Methods for Evaluating Perceived Physical Effort," *Journal of Applied Psychology*, Vol. 65, No. 6, 1980, pp. 672–679.

Hollnagel, E., "Task Analysis: Why, What, and How," in G. Salvendy, ed., *Handbook of Human Factors and Ergonomics*, 3rd ed., Hoboken, N.J.: John Wiley & Sons, 2006, pp. 373–383.

Jackson, A. S., "Preemployment Physical Evaluation," *Exercise and Sport Sciences Reviews*, Vol. 22, 1994, pp. 53–90.

Jackson, Andrew S., "Types of Physical Performance Tests," in Stefan Constable and Barbara Palmer, eds., *The Process of Physical Fitness Standards Development*, Wright-Patterson Air Force Base, Ohio: Human Systems Information Analysis Center, 2000, pp. 101–138.

Kalns, J., J. Baskin, A. Reinert, D. Michael, A. Santos, S. Daugherty, and J. K. Wright, Predicting Success in the Tactical Air Combat Party Training Pipeline. *Military Medicine*, Vol. 176, No. 4, 2011, pp. 431–437.

Karhu, O., P. Kansi, and I. Kuorinka, "Correcting Working Postures in Industry: A Practical Method for Analysis," *Applied Ergonomics*, Vol. 8, No. 4, 1977, pp. 199–201.

Kelly, G. A., *The Psychology of Personal Constructs*, New York: Norton, 1955.

Kiesel, K., P. Plisky, and P. Kersey, "Functional Movement Test Score as a Predictor of Time-loss During a Professional Football Team's Pre-Season," *Medicine and Science in Sports and Exercise*, Vol. 40, No. 5, 2008, pp. S234–S234.

Knapik, Joseph, Philip Ang, Katy Reynolds, and Bruce Jones, "Physical Fitness, Age, and Injury Incidence in Infantry Soldiers," *Journal of Occupational Medicine*, Vol. 35, No. 6, 1993, pp. 598–603.

Knapik, J. J., E. A. Harman, R. A. Steelman, and B. S. Graham, "A Systematic Review of the Effects of Physical Training on Load Carriage Performance," *Journal of Strength and Conditioning Research*, Vol. 26, No. 2, 2012, pp. 585–597.

Knapik, J. J., W. Harper, and H. P. Crowell, "Physiological Factors in Stretcher Carriage Performance," *European Journal of Applied Physiology and Occupational Physiology*, Vol. 79, No. 5, 1999, pp. 409–413.

Knapik, J. J., K. L. Reynolds, and E. Harman, "Soldier Load Carriage: Historical, Physiological, Biomechanical, and Medical Aspects," *Military Medicine*, Vol. 169, No. 1, 2004, pp. 45–56.

Knapik, Joseph J., Bruce H. Jones, Marilyn A. Sharp, Salima Darakjy, Sarah Jones, Keith G. Hauret, and Gene Piskator, *The Case for Pre-Enlistment Physical Fitness Testing: Research and Recommendations*, Aberdeen Proving Ground, Md.: U.S. Army Center for Health Promotion and Preventive Medicine, 12-HF-OIQ9D-04, 2004.

Latham, G. P., and K. N. Wexley, *Increasing Productivity Through Performance Appraisal*, 2nd ed., Reading, Mass.: Addison-Wesley, 1993.

Legaz-Arrese, A., D. Munguia-Izquierdo, L. E. Carranza-Garcia, and C. G. Torres-Davilla, "Validity of the Wingate Anaerobic Test for the Evaluation of Elite Runners," *The Journal of Strength and Conditioning Research*, Vol. 25, No. 3, 2011, p. 819.

Lopez, F. M., *The Threshold Traits Analysis*, Port Washington, N.Y.: Lopez & Associates, 1986.

Lievens, F., and B. De Soete, "Simulations," in N. Schmitt, ed., *The Oxford Handbook of Personnel Assessment and Selection*, New York: OUP USA, 2012, pp. 383–410.

Marras, W. S., "Occupational Biomechanics," in W. Karwowski and W. S. Marras, eds., *Occupational Ergonomics: Principles of Work Design*, Boca Raton, Fl.: CRC Press, 2003, pp. 10-1–10-38.

Mattila, M., and M. Vilkki, "OWAS Methods," in W. Karwowski and W. S. Marras, eds., *Occupational Ergonomics: Principles of Work Design*, Boca Raton, Fl.: CRC Press, 2003, pp. 26-1–26-13.

McAtamney, L., and E. N. Corlett, "RULA: A Survey Method for the Investigation of Work-Related Upper Limb Disorders," *Applied Ergonomics*, Vol. 24, No. 2, 1993, pp. 91–99.

McKeon, P. O., and J. Hertel, "Systematic Review of Postural Control and Lateral Ankle Instability, Part I: Can Deficits Be Detected with Instrumented Testing?" *Journal of Athletic Training*, Vol. 43, No. 3, 2008, p. 293.

McPhail, S. Morton, "Development of Validation Evidence," in S. Morton McPhail, ed., *Alternative Validation Strategies: Developing New and Leveraging Existing Validity Evidence*, San Francisco, Calif..: Jossey-Bass, 2007, pp. 1–28.

McPhail, S. Morton, and Damian J. Stelly, "Validation Strategies," in John C. Scott and Douglas H. Reynolds, eds., *Handbook of Workplace Assessment: Evidence-Based Practices for Selecting and Developing Organizational Talent*, San Francisco, Calif..: Jossey-Bass, 2010, pp. 671–710.

Mello, R. P., A. I. Damokosh, K. L. Reynolds, C. E. Witt, and J. A. Vogel, *The Physiological Determinants of Load Bearing Performance at Different March Distances*, Technical Report T15-88, Natick, Mass.: U.S. Army Research Institute of Environmental Medicine, 1988.

O'Connor, F. G., P. A. Deuster, J. Davis, C. G. Pappas, and J. J. Knapik, J. J., "Functional Movement Screening: Predicting Injuries in Officer Candidates," *Medicine and Science in Sports and Exercise*, Vol. 43, No. 12, 2011, pp. 2224–2230.

Pate, R. R., M. L. Burgess, J. A. Woods, J. G. Ross, and T. Baumgartner, "Validity of Field-Tests of Upper-Body Muscular Strength," *Research Quarterly for Exercise and Sport*, Vol. 64, No. 1, 1993, pp. 17–24.

Quesada, P. M., "Video-Based Measurements of Human Movement," in W. Karwowski and W. S. Marras, eds., *Occupational Ergonomics: Principles of Work Design*, Boca Raton, Fl..: CRC Press, 2003, pp. 32-1–32-16.

Rayson, M., D. E. Holliman, R. V. Nevola, and C. L. Birch, *Physical Selection Standards for the British Army: Phase 5 Validation*, Technical Report, PLSD/CHS5/CR96/021, Farnsborough, UK: Center for Human Sciences, 1996.

Rayson, M. P., "Job Analysis," in S. Constable and B. Palmer, eds., *The Process of Physical Fitness Standards Development*, Wright-Patterson Air Force Base, Ohio: Human Systems Information Analysis Center, HSIAC-SOAR-2000-001, December 2000, pp. 67–98.

Rayson, M. P., and D. E. Holliman, *Physical Selection Standards for the British Army: Phase 4 Predictors of Task Performance in Trained Soldiers*, Technical Report, DRA/CHS/PHYS/CR95/017, Farnsborough, UK: Defence Research Agency, 1995.

Rayson, M. P., D. G. Bell, D. E. Holliman, M. Llewellyn, R. V. Nevola, and R. L. Bell, *Physical Selection Standards for the British Army Phases 1 and 2*, Technical Report, DRA/CHS/HS2/2007, Farnsborough, UK: Army Personnel Research Establishment, 1994.

Rayson, M., D. Holliman, and A. Belyavin A., Development of Physical Selection Procedures for the British Army. Phase 2: Relationship Between Physical Performance Tests and Criterion Tasks," *Ergonomics,* Vol. 43, No. 1, 2000, pp. 73–105.

Rhea, M. R., B. A. Alvar, and R. Gray, R., "Physical Fitness and Job Performance of Firefighters," *Journal of Strength and Conditioning Research*, Vol. 18, No. 2, 2004, pp. 348–352.

Rice, V. J. B., M. A. Sharp, W. J. Tharion, and T. L. Williamson, "The Effects of Gender, Team Size, and a Shoulder Harness on a Stretcher-Carry Task and Post-Carry Performance. Part I. A Simulated Carry from a Remote Site," *International Journal of Industrial Ergonomics*, Vol. 18, No. 1, 1996, pp. 27–40.

Richmond, V. L., M. P. Rayson, D. M. Wilkinson, J. M. Carter, S. D. Blacker, A. Nevill, J. Du Ross, and S. Moore, "Development of an Operational Fitness Test for the Royal Air Force," *Ergonomics*, Vol. 51, No. 6, 2008, pp. 935–946.

Romines, Reggie, "Justification for Annual Medical Evaluations for Fire Fighters," Executive Leadership Course, applied research project submitted to the National Fire Academy, December 1998, p. 10.

Schmidt, F. L., and Hunter, J. E., *Methods of Meta-Analysis: Correcting Error and Bias in Research Findings*, Sage Publications, Inc., 2004.

Schmitt, Neal and Ruchi Sinha, "Validation Support for Selection Procedures," in Sheldon Zedeck, ed., *APA Handbook of Industrial and Organization Psychology, Vol. 2: Selecting and Developing Members of the Organization*, Washington, D.C.: American Psychological Association, 2011, pp. 399–420.

Serra, Consol, Mari Cruz Rodriguez, George L. Delclos, Manel Plana, Luis I. Gómez López, and Fernando G. Benavides, "Criteria and Methods Used for the Assessment of Fitness for Work: A Systematic Review," *Journal of Occupational and Environmental Medicine*, Vol. 64, No. 5, 2007, pp. 304–312.

Shephard, R. J., "Assessment of Occupational Fitness in the Context of Human Rights Legislation," *Canadian Journal of Sport Science*, Vol. 15, No. 2, 1990, pp. 89–95.

Sheppard, J. M., and W. B. Young, "Agility Literature Review: Classifications, Training and Testing," *Journal of Sports Sciences*, Vol. 24, No. 9, 2006, pp. 919–932.

Shrout, P. E., and J. L. Fleiss, "Intraclass Correlations: Uses in Assessing Rater Reliability," *Psychological Bulletin*, Vol. 86, No. 2, 1979, p. 420.

Sporis, G., I. Jukic, L. Milanovic, and V. Vucetic, "Reliability and Factorial Validity of Agility Tests for Soccer Players," *Journal of Strength and Conditioning Research*, Vol. 24, No. 3, 2010, pp. 679–686.

Stanish, Heidi I., and Terry M. Campagna, "Prediction of Performance on the RCMP Physical Ability Requirement Evaluation," *Journal of Occupational and Environmental Medicine*, Vol. 41, No. 8, August 1999, pp. 669–677.

Steel, P. D., and J. D. Kammeyer-Mueller, "Comparing Meta-Analytic Moderator Estimation Techniques Under Realistic Conditions," *Journal of Applied Psychology*, Vol. 87, No. 1, 2002, pp. 96–111.

Theologus, G. C., and E. A. Fleishman, "Development of a Taxonomy of Human Performance: Validation of the Ability Scales for Classifying Human Tasks," *JSAS Catalog of Selected Documents in Psychology*, Vol. 3, Ms. No. 326, 1973, p. 29.

Theologus, G. C., T. Romanshko, and E. A. Fleishman, "Development of a Taxonomy of Human Performance: A Feasibility Study of Ability Dimensions for Classifying Human Tasks," *JSAS Catalog of Selected Documents in Psychology*, Vol. 3, Ms. No. 321, 1973, pp. 25–26.

Tipton, M. J., G. S. Milligan, and T. J. Reilly, "Physiological Employment Standards I. Occupational Fitness Standards: Objectively Subjective?" *European Journal of Applied Physiology*, 2012, pp. 1–12.

Truxillo, Donald M., "Setting Cutoff Scores for Personnel Selections Tests: Issues, Illustrations, and Recommendations," *Human Performance*, Vol. 9, No. 3, 1996, pp. 275–295.

U.S. Air Force, "Special Operations Weather Team," fact sheet, May 26, 2010a. As of October 28, 2014:
http://www.af.mil/AboutUs/FactSheets/Display/tabid/224/Article/104529/special-operations-weather-team.aspx

U.S. Air Force, "Pararescue," fact sheet, June 17, 2010b. As of October 28, 2014:
http://www.af.mil/AboutUs/FactSheets/Display/tabid/224/Article/104515/pararescue.aspx

U.S. Air Force, "Combat Controllers," fact sheet, August 18th, 2010c. As of October 28, 2014:
http://www.af.mil/AboutUs/FactSheets/Display/tabid/224/Article/104592/combat-controllers.aspx

van Dijk, J. "Common Military Task: Marching," in *Optimizing Operational Physical Fitness* (Vol. TR-HFM-080), NATO Research and Technology Organisation, 2009.

Vanderburgh, P. M., "Occupational Relevance and Body Mass Bias in Military Physical Fitness Tests," *Medicine and Science in Sports and Exercise*, Vol. 40, No. 8, 2008, pp. 1538–1545.

Verdijk, L. B., L. Van Loon, K. Meijer, and H. H. C. M. Savelberg, "One-Repetition Maximum Strength Test Represents a Valid Means to Assess Leg Strength in Vivo in Humans," *Journal of Sports Sciences*, Vol. 27, No. 1, 2009, pp. 59–68.

Vickers, R. R., Jr., *Running Performance as an Indicator of V02 Max: A Replication of Distance Effects*, Technical Report No. 01–24. San Diego, Calif.: Naval Health Research Center, 2001.

Vickers, R. R., Jr., *Modeling Run Test Validity: A Meta-Analytic Approach*, Technical Report No. 02-27, San Diego, Calif.: Naval Health Research Center, 2002.

von Restortf, W., "Physical Fitness of Young Women: Carrying Simulated Patients," *Ergonomics*, Vol. 43, No. 6, 2000, pp. 728–743.

Walker, T. B., L. Lennemann, C. Mauzy, J. N. McGregor, and M. F. Zupan, *Physiological and Psychological Characteristics of Successful Combat Controller Trainees*, DTIC Document, 2010.

Walker, T. B., L. Lennemann, M. F. Zupan, V. Anderson, and W. Lyons, *Adaptations to a New Physical Training Program in the Combat Controller Training Pipeline*, DTIC Document, 2010.

Zedeck, S., J. C. Hogan, W. F. Cascio, I. L., Goldsten, R. S. Barrett, and J. Outtz, *Development of Tests for Entry Level Firefighters*, Pleasant Hill, Calif.: CORE Corporation, 1988.